D1796173

The Chemical Industry

The Chemical Industry

PAST AND PRESENT

TREVOR ILLTYD WILLIAMS

M.A. B.SC. D.PHIL. F.R.I.C.

with a new Foreword
by
The Author

Republished EP Publishing Limited 1972
First published 1953

Republished 1972 EP Publishing Limited,
East Ardsley, Wakefield
Yorkshire, England,
by kind permission of the copyright holder

© TREVOR ILLTYD WILLIAMS

This edition offset by kind
permission of the original publisher
Penguin Books Ltd.

ISBN 0 85409 813 5

Please address all enquiries to
EP Publishing Limited
(address as above)

Reprinted in Great Britain by
Scolar Press Limited, Menston, Yorkshire

For Sylvia

Contents

Foreword

to Open University reprint

This work was written some twenty years ago and was designed to give the general reader a concise account of the history of the chemical industry, of its organization at that time, and of the nature of its products. It concluded with some cautious speculation about the future of the industry.

In the main, the predictions made in the final chapter have been fulfilled. The industry has expanded enormously; the expansion has been particularly great in the polymer field; petroleum has virtually replaced coal as a raw material for industrial organic chemicals. No particular merit attaches to such prescience, however, for it represented no more than a logical extrapolation of trends already apparent, at least within the industry.

To do justice to all the changes that have taken place during the last twenty years would, however, demand some considerable extension of the book, for many novel products and processes have appeared and there has been much change of emphasis within the industry. Some explanation of its present reprinting in its original form is, therefore, called for. The reason lies in the fact that it is now reissued not for the benefit of the general reader but for students of science and technology in the Open University. Where history ends and contemporary affairs begins is a matter for argument, but I think it fair to say that, within its modest compass, this work still adequately records the development of the chemical industry up to about the middle of this century. For the historian, this is late enough.

Many will, of course, wish to read more widely and deeply, and to facilitate this a Reading List is given at the end of the book. This was compiled in 1952, and since that time several important works have appeared with which the serious student of the subject should be familiar. The following are particularly noteworthy:

The Chemical Industry during the Nineteenth Century, by L. F. Haber. Clarendon Press, 1958 (reprinted with corrections 1969).

The Chemical Industry 1900–1930, by L. F. Haber. Clarendon Press, 1971.

Imperial Chemical Industries: a History. Vol. I: the Fore-runners 1870–1926, by W. J. Reader. Oxford University Press, 1970. (Vol. II in preparation)

A History of the Modern British Chemical Industry, by D. W. F. Hardie and J. Davidson Pratt. Pergamon Press, 1966.

The Story of the Chemical Industry in Basle. Ciba Ltd., 1959.

Enterprise in Soap and Chemicals; Joseph Crosfield & Sons Ltd. 1815–1965, by A. E. Musson. Manchester University Press, 1965.

June 1972   TREVOR I. WILLIAMS

Preface

T HE chemical industry differs from almost all others in two important ways. First, it does not provide a single or a comparatively small number of products and services as do, for example, the textile, shipbuilding, and mining industries. Its products are numbered in thousands, requiring very widely differing methods and equipment for their production. Second, although the operations of the chemical industry affect almost every phase of daily life its products are for the most part absorbed by other branches of industry and therefore are often quite unknown to the general public.

This book is an attempt to describe simply and briefly how the chemical industry is organized, what it makes, and how its products are used. A substantial part of the book is historical, for it was felt that without an understanding of the way in which the industry has evolved it is impossible to understand why its modern organization is so exceedingly complex. Although the emphasis is on the British industry there is extensive reference to processes of foreign origin, for chemistry, like science as a whole, has always been an essentially international enterprise. So far as possible technical terms have been avoided or explained, but as the whole industry is an example *par excellence* of applied science, and many of its products have no popular names, a good deal of unfamiliar terminology is unavoidable.

Any brief survey of so complex and far-reaching an organization must be to some extent subjective, for different people will necessarily make a somewhat different choice of what to include and what to omit, what to emphasize and what to let pass without comment. In this book I have sought only to describe the chemical industry as it presents itself to me.

It is a pleasure to express my thanks to a number of friends and colleagues who have generously given their advice and encouragement. Sir Wallace Akers was kind enough to read the whole of the manuscript and, from his deep knowledge of the subject, enabled me to correct a number of errors and to effect some improvements in the style. Mr Henry Maxwell offered, when the work was being planned, some pertinent advice on the arrangement of material. Mr George White and Dr J. Gordon Cook most generously advised me on the chapter dealing with dye-stuffs and pigments. Finally, I must record my appreciation of the unfailing care taken by Mrs A. Swinnerton in all the secretarial work involved.

T. I. W.

Oxford, 1952.

PART I: PAST

CHAPTER ONE

The Beginning

THE chemical industry is one of the most ancient of all industries, for the very earliest written records show it then to have been firmly established and prehistoric remains prove an undoubted familiarity with certain simple chemical processes. It is nevertheless very difficult to recount the early history of industrial chemistry. The earliest records are, as one would expect, very far from complete and those we have are often hard to decipher because, as valued trade secrets, handed down from father to son, they were often deliberately written in terms not intelligible to the uninitiated.

Our knowledge of the earliest phases of the chemical industry comes from Egypt. This is not necessarily an indication that Egypt was the cradle of the chemical industry, a centre from which knowledge flowed outwards, gaining in extent by local additions as it did so until it covered the whole of the ancient world: it is rather a reflection of the fact that western scholars have concentrated the weight of their attention on Egyptian archaeology to the detriment of studies of the early histories of scarcely less ancient civilizations, such as those of China. It may well be that in China and one or two other centres chemical processes similar to those of ancient Egypt were carried on contemporaneously or even at an earlier date. In writing of the growth of a primitive chemical industry in Egypt, however, we are on fairly sure ground, for there is a wealth of documentary and sound archaeological evidence to support at least the main points in the story. The smelting of metals was certainly almost, if not quite, the earliest of chemical processes discovered by man; undoubtedly it was the most important, for, by greatly increasing his power of triumphing over his environment, it promoted the creation of the secure and settled way of life essential for the development of other phases of civilization. With knowledge of the smelting of metals came knowledge of the properties of their chemical compounds.

It is very likely that the first metal known was gold, for this occurs in many parts of the world in its native state, requiring no treatment whatsoever before being worked. Archaeological evidence shows, however, that copper, which occurs native in only a few parts of the world, was known at almost as early a date as gold; in Egypt, indeed, it may have been known before gold. This is a point of the greatest interest, for copper can be extracted from its ores only by means of the chemical changes comprised in the process of smelting. The Egyptians probably obtained most of their copper from the ore known as malachite, the basic carbonate of the metal, used also as a green pigment.

Iron, too, was known at a very early date, at least as early as 3400 B.C. It was then, however, available only in very small quantities and was no doubt obtained in the native state from meteorites. The discovery of the smelting of iron ores was made much later. Not until 2000 B.C. was iron common in Egypt; most of it was then probably imported from the Hittites; by 1400 B.C.

the Iron Age was firmly established and the Assyrians were making extensive use of the metal. Lead is another metal which was known at an early date – a statuette in the British Museum dates from the First Dynasty – but did not come into general use until very much later. At an early, but undetermined date, processes for preparing the oxide of lead (litharge) and its acetate (white lead) were discovered; both these substances are still widely used as pigments. Bronzes, alloys of tin and copper, were also in very early use. A specimen from Ur has been dated as of 3500 B.C.; bronze is known to have been in use in Egypt no more than a century later. It is very interesting to realize that even in those remote times nations sent almost as far afield for their raw materials as we do to-day; far farther afield, indeed, if allowance is made for the much greater danger and difficulty of travel. Egypt, as we have already seen, established the importation of iron from Assyria some four thousand years ago; only a little less ancient, according to some historians, was her importation of tin from Britain. These islands became known in the ancient world as the 'Islands of the Kassiterides' from the Greek word, *kassiteros*, for tin. Silver is another metal widely known in the earliest times of which we have records, and originally more highly prized than gold. Quicksilver, or mercury, was known at least as early as 2500 B.C. Brass, however, an alloy of copper and zinc, came into use very much later; Plato's *orichalcum* was probably a type of brass.

The development of processes for extracting metals from their ores is only a partial solution of the problem of turning them to practical use. No less important is the development of methods of working them; craftsmanship is no less essential than chemistry. Similar conditions apply in other fields of human effort; many traditional crafts involve the use, at some time or another, of chemical processes. For example, leather has been used for many purposes since the earliest times of which we have knowledge, but our own experience informs us that unless hide is carefully treated after being stripped from the beast it rapidly becomes hard and useless. The value of alum for the preparing of leather was discovered very early in man's history. By a natural accident this chemical is available in a state of high purity from many sources, and it is among the oldest products of the chemical industry. Chemical analyses of ancient leather goods show the presence of aluminium – the chief metal present in alum – in quantities sufficiently high to make it very probable that this chemical was used in their preparation. Homer's Iliad, for example, describes the preparation of leather, and leather of this period has been proved to contain a good deal of aluminium.

The first written reference to alum is in the famous Ebers Papyrus, which dates from approximately 1500 B.C. This papyrus is essentially a collection of technical recipes and the style in which it is written makes it clear that it is not an original work but a compilation of knowledge already well established. We can, therefore, be sure that alum was used in Egypt well before 1500 B.C., so that it must have been in continuous use for roughly four thousand years. It was valued for many purposes other than the treating of leather. Its use as a styptic, for example, was well appreciated. More important still was its use in dyeing. Many natural dyes will not adhere firmly to fibres, such as those of cotton, unless the latter have first been treated with substances known as mordants; of these alum is among the most useful. Alum used for this purpose must, however, be of high purity; as much of the natural mineral contains traces of iron, which would stain any fabric dipped in it, it is reasonable to suppose that the Egyptians were familiar with the process of recrystallizing alum, and other soluble salts, from water in order to purify it.

The preparation of the dyes with which the alum mordant was used also entailed some skill in chemical manipulation. Much dyeing was no doubt done by means of extracts of local herbs and lichens, just as it is to-day in the Outer Hebrides for the making of Harris tweed; as one would expect, our knowledge of the specific plants used for this purpose is limited. Of more brilliant and powerful dyes, however, we have a much better knowledge, for their use has, in some cases, persisted up to our own times. Indigo, for example, was in use at least four thousand years ago and the discovery, within the last century, of means of synthesizing indigo commercially has still not entirely killed the natural indigo industry, though it has inevitably dealt it a very heavy blow. Another famous dye of the ancient world was Tyrian Purple, extracted from a species of small Mediterranean mollusc. It is interesting to note that, despite their different colours and their very different natural sources, recent chemical investigation has shown these two ancient dyes to be very closely related; Tyrian Purple is merely indigo into whose structure a little bromine has been introduced.

The making of glazed pottery, and somewhat later of glass ware, entailed the preparation of several different chemicals and some knowledge of their properties and uses. Even in the First Dynasty blue glazes for pottery were made with the help of copper salts; a pale green glass bead – its colour probably due to the accidental rather than deliberate addition of iron – dates from the same period. By roasting malachite with sand and lime a blue pigment known as Egyptian Blue was made. This was used, with soda, for making a blue glaze on faience.

Soda, obtained under the name of natron from various natural deposits, was used for the making of glass. Glass manufacture on a considerable scale was established by 1370 B.C. but undoubtedly it existed for many centuries before this. Some early examples of blue glass owe their colour to the presence of cobalt, but this was probably fortuitous and it is very doubtful whether the Egyptians had any specific knowledge of the compounds of this metal.

There is some evidence that even at the earliest times of which we have knowledge the process of causticizing soda by means of lime – made by roasting limestone – was known. In the Sumerian period a crude soap was made from oil and alkali; this would have been possible only if caustic alkali had been available. Potash – or 'vegetable alkali' – made by the burning of plant materials was also apparently well known to the Egyptians, for analyses of some of the earliest glass vessels show the presence of potash in quantities greater than can readily be accounted for as an accidental impurity. Fuller's earth, a valuable cleansing agent, occurs widely and was well known throughout the Near East at an early date.

Nitre or saltpetre (potassium nitrate) was also certainly known in most ancient times and may have been used for making nitric acid, but original accounts are often difficult to understand owing to the confusion which existed between nitre and natron (soda). Nitre was obtained from two sources: from certain natural deposits, especially in Babylonia, and from the incrustation which formed on decaying organic material such as the sweepings from stables. Sal ammoniac (ammonium chloride) was made by heating camels' dung and redistilling the soot, but it is uncertain when this process was discovered. It is possible that the Egyptian *nitrum* mentioned by Pliny was in fact crude sal ammoniac. Common salt (sodium chloride), obtained in a fairly pure state by the solar evaporation of sea-water in hot countries, has been a basic article of trade from the very earliest times and in parts of Asia and Africa is still as readily negotiable as money. Early accounts are difficult to disentangle; even as late as the seventeenth century Boyle defined salt no more clearly than by saying it was a substance soluble in water and having a strong taste either pleasant or unpleasant.

It is said, though the evidence is not strong, that the Egyptians were familiar with elementary yellow sulphur and may even have made sulphuric acid from it. It is certain, however, that sulphur exists in many natural deposits around the Red Sea, probably formed by bacterial decomposition of sulphates in former lakes. It is interesting to note that this process occurs to-day in certain brackish lakes in Cyrenaica and in the Sudan, where the crude sulphur is annually harvested by the local Arabs, and that the possibility of utilizing it industrially during the recent world shortage of sulphur was very closely investigated in laboratories in Britain.

Certain organic products also, apart from dyes, have been well known for thousands of years. Fermentation processes for making alcohol, primarily for the production of intoxicating liquors, seem to be almost as old as mankind. It was also well known that the fermentation might proceed a stage further, the alcohol being converted into acetic acid, from which may be derived, among other products, the lead salt, used as a white pigment. Another organic substance available to the ancient world in a very pure form was sugar, refined by crystallization from water. Various oils, fats, and waxes were also well known.

We see, therefore, that it is not possible to trace the story of the chemical industry to its ultimate source. At the very earliest times of which we have a record a considerable number of chemical processes had been discovered and many chemicals with which we are familiar to-day were even then in daily use. Among the metals available were copper, silver, gold, iron, tin, lead, and mercury. Among the salts – often very confusingly described – in general use were common salt, alum, soda, nitre, sal ammoniac, and potash. Sulphur, limestone, lime, malachite, quartz (sand), and fuller's earth, were familiar minerals. Among organic substances in use were various vegetable dyes, sugar, products of alcoholic and acetic fermentation, oils, fats, and waxes.

For many centuries progress was very slow, the main development being in craftsmanship rather than in the processes supplying the raw materials. Greece and Rome, with civilizations surpassing anything then existing, gave due honour to the craftsman, yet chemical knowledge and chemical processes still remained almost stationary. The world was on the whole content to exploit more and more fully the materials which lay ready to hand rather than to seek for new ones. It is not unfair to claim that at the dawn of the Christian era, and indeed for some centuries afterwards, chemistry did not greatly differ basically from its state of two thousand years earlier.

Not until a new outlook – engendered in part by the always powerful motive of avarice – was evolved did any great progress in chemistry occur. Men at last began to seek an insight into the nature of chemical processes instead of merely accepting them as convenient empirical means to achieve practical ends. For this change we can see two main reasons, although it is, of course, an oversimplification to suppose that so complex a change could be attributed to only two factors.

On the one hand, the process of civilization and the development of strong and lasting empires created the conditions necessary for the formation of a class of scholars – as opposed to a priestly caste – with the time and means to occupy themselves with purely intellectual activities instead of having to concentrate their attention upon the day-to-day business of finding the necessities of life. For the most part, especially in Greece and Rome, scholars gave their attention to essentially artistic studies – poetry, drama, art, sculpture, and so on – which accordingly flourished exceedingly. Some few, however, applied their minds to a study of the nature of the world in which they found themselves. In this, however, the Greeks and Romans found themselves severely handicapped by the conventions of their own civilizations. They were, for the most part, obliged to base their deductions merely upon the results of observations of natural phenomena – often relying upon the inaccurate reports of others – and not upon the practical experiments which in recent times have paid such a rich dividend. The making of experiments was regarded as a manual occupation more fit for slaves than for free men.

On the other hand, the simple chemical changes known to the ancient world – for example the conversion of bright green lumps of malachite into shining plates of polished copper – suggested, not unreasonably, the possibility of other transformations, some of which we now know to be impossible to effect by purely chemical means. In particular, it suggested that base metals might be transmuted into gold, a metal which, from the earliest times and among all nations, has had the high and largely artificial value which it has to-day. Actuated by their urgent wish to prepare the Philosopher's Stone – a semi-mystical object regarded as the penultimate step in the manufacturing of gold – the alchemists, as the early chemists pursuing this particular line of thought are called, inevitably acquired an extensive store of miscellaneous chemical information. They subjected to every kind of chemical manipulation known to them every kind of chemical substance which they could acquire. These operations, unfortunately described for the most part in highly mystical terms, had several important results, although the creation of gold was not one of them. Substances already in common industrial use were so closely investigated that their properties were in many instances clearly and unequivocally defined. By various chemical reactions, or by study of hitherto neglected minerals, quite new substances were identified. No less important was the evolution of simple chemical apparatus such as stills, furnaces, crucibles, mortars, and so on.

We have already seen that the reluctance of the Greeks and Romans to carry out experiments severely limited their progress in chemistry. Nevertheless they did, by thinking deeply about the natural phenomena with which they were familiar, make some contributions to chemistry as a science rather than as a purely empirical art. Democritus, for example, conceived a simple atomic theory as long ago as the fourth century B.C. Only a little later Aristotle put forward the idea that all matter consisted of combinations, in varying proportions, of four elements – air, earth, fire, and water. Erroneous though it was, being based on mere speculation, Aristotle's theory has a double significance. On the one hand it illustrates an early attempt to account logically for the very different properties of the substances we encounter in daily life; on the other it influenced all chemical thought – and thus the chemical industry – for no less than two thousand years. We shall see later that Aristotle's elements were altogether different from elements as we regard them to-day and as first defined by Robert Boyle in the seventeenth century.

On the whole the Greek and Roman contributions to chemistry were very small, though notable for making the first theoretical approach. For the first outstanding achievements in chemistry we must turn to a very different but no less talented race – the Arabs – who were most probably inspired primarily by a philosophical school established in the early centuries of the Christian era in Egypt and Byzantium. The foundations on which the Arabs had to build were, however, very slight and it is largely to their efforts that we owe such chemical knowledge as existed in the tenth century. Their main inheritance from Egypt and Byzantium was a variety of apparatus specially designed for carrying out such chemical manipulations as distillation, sublimation, grinding, roasting, and so on. This apparatus was constructed mostly of pottery, iron, or wood. Not until comparatively recent times was glass laboratory ware introduced; the influence of this on the development of chemistry was immense and even now not at all generally appreciated. The main secret of the Arabs' success, however, was their readiness to carry out careful experiments to supplement the knowledge gained by observation of natural phenomena.

The Arabs discovered many new chemicals, the most important of them being the mineral acids, to-day of immense industrial importance; it is possible that both sulphuric and nitric acids were known to the Egyptians, but if so they certainly never studied them closely nor appreciated their importance. No less important, they gave clear descriptions of chemicals already known. The most important Arab contributions are included in works attributed to Geber, who lived in the eighth century. It is very probable, however, that many of these works were in fact written by other industrious chemists whose very names are now unknown. Other outstanding Arab chemists were Rhazes and Avicenna.

Avicenna died in 1037, bringing our story up to the end of the first millennium of the Christian era. From this point onward we are on very much surer ground, for we can rely more and more on authentic written records. The following five centuries or so we may term the golden age of alchemy – even though gold, the ultimate goal of the alchemists, eluded them. The force of Arab chemistry was spent but its accumulated knowledge slowly found its way to Europe and was extended by such men as Albertus Magnus in the twelfth century and Roger Bacon and Raymond Lully in the thirteenth. A very shrewd account of contemporary alchemy is to be found in Chaucer's *Chanouns Yemannes Tale*. The chief tools of the alchemists were the crucible and the furnace:

> We blondren ever and pouren in the fyr
> And for al that we fayle of our desire.

Contemporary prints show that the alchemist's den had much in common with the blacksmith's forge. Chaucer quotes at some length the materials used by the alchemists of his day, making it clear that a considerable range of chemicals was then available:

> Unslékked lym, chalk, and gleyre of an ey,
> Poudres diverse, asshes, dong, pisse, and cley,
> Cered pokets, sal petre, vitriole;
> And divers fyres maad of wode and cole;
> Sal tartre, alkaly, and sal preparat,
> And bombust materes and coagulat,
> Cley maad with hors or mannes heer, and oile
> Of tartre, alum, glas, berm, wort, and argoile;
> Re(s)algar, and our materes enbibing
> And of our silver citrinacioun.

From Norton's *Ordinall of Alchemy* we can glean more information about the materials then in use. They include antimony, arsenic, marcasite, and sulphuric acid. James IV of Scotland was very fond of dabbling in alchemy and employed a foreign alchemist – probably a Florentine – named Damian. Extant accounts show that Damian used large quantities of 'aqua vite' which was almost certainly no more than a type of whisky. No doubt alchemists, like their modern counterparts, were very appreciative of the subtle virtues of this as a beverage; equally, however, they were intrigued by its apparent harmonious blending of the opposing elements of fire and water. However he divided his interests between these two aspects of whisky, Damian required large quantities of it. He also handled such substances as cinnabar, litharge, white lead, tin, verdigris, orpiment, mercury, sal ammoniac, and saltpetre.

The alchemists indulged in theoretical speculations, though these were essentially mystical in nature and not derived from experiment or experience. They evolved the idea of two primary elements – mercury and sulphur – from the varying combinations of which all other metals were formed. It is clear, however, that, at least in the higher ranks of the mystery, these were not regarded as the ordinary mercury and sulphur we encounter in ordinary practice and both of which have been available in very pure forms since ancient times. The two elements were clearly regarded as some idealized forms of sulphur and mercury. They also conceived the idea that metals would 'grow' in the same sort of way as plants grow in the soil; in the later age of alchemy the alchemists begged gold from their patrons with the promise that 'of a single pound we can make tweye'.

Although much genuine, if misguided, work was done, trickery was rampant. The Canon's Yeoman, for example, describes the ancient trick of hiding the substance supposed to be produced in an experiment inside the hollow shaft of the spoon with which the mixture is stirred. Another trick was to conceal a metal in the coals of the fire.

> This false canoun (the foule feend him fetche!)
> Out of his bosom took a false cole
> In which ful subtily was made an hole
> And therein was put of silver metal.

Despite all the trickery, however, and all the setbacks such as explosions which 'be of so gret violence Our walles may not make them resistence', the alchemist retained his faith in ultimate success.

Although they failed in their main objective, the knowledge gained by the alchemists could be put to practical use in ways other than in the futile quest after synthetic gold and stimulated the growth of chemical industry. The Renaissance was marked by a great increase in metallurgy and this led to a need for further simple chemical manufactures. Acids were required, for example, in considerable quantities for the separation of silver and gold from Bohemian ores. In 1556 Agricola published his *De Re Metallica*, a large and essentially practical work on mining and metallurgy. Shortly afterwards Libavius published a work called *Chymia* which has been called the first chemical textbook, although it was still deeply imbued with the mystical lore of alchemy.

Libavius was a follower of a remarkable man generally known as Paracelsus – his full name was Philip Aureolus Theophrastus Bombast von Hohenheim; from him the English language has gained the words bombast and bombastic. Though Paracelsus had some unusual ideas, which included a belief that all matter was animated by spirits, he performed a very valuable service to chemistry by directing attention away from the futile aim of synthesizing gold, towards a study of the therapeutic possibilities of the various substances which had become familiar. From this new approach derived, for example, Glauber's valuable discovery of the purgative properties of sodium sulphate (Glauber's salt). It may be noted in passing that to-day the medical profession is dependent on the chemical industry in almost every phase of its work.

Through all these centuries, however, chemistry remained an empirical art, followed either for such practical ends as the carrying out of metallurgical processes or with the hope of synthesizing gold; many facts were clearly understood but scarcely any chemical theory existed. By the seventeenth century, however, the time was ripe for a great change. After centuries of unquestioning acceptance of so-called facts which had never been confirmed, the western world was in an inquiring frame of mind. In 1610 Francis Bacon, in his *Novum Organum*, stressed the importance of first collecting all available facts, either by observation or by experiment, and then trying to formulate theories to account for them; hitherto men had fallen into the error of formulating their theories first and then trying to adapt the known facts to them. Bacon had little use for facts as such, but only when they could be fitted into a common framework; he believed, too, that knowledge should be applied to allow man to master nature. This latter point is of the greatest importance, for the chemical industry depends for its very existence upon man's ability to apply natural forces to his own benefit.

Bacon's teaching did not lose its momentum with his death in 1626, for in the following year was born Robert Boyle, a man well fitted to succeed him. Boyle may truly be called the father of modern chemistry. He disputed the four-element theory which had survived since the time of Aristotle and defined an element in essentially the same way as we do to-day, namely as a substance which cannot by any means be broken down into simpler ones. He challenged the mysticism and obscure writing of the alchemists – so often a cloak for ignorance – and expressed himself freely and lucidly. He stated bluntly that 'he that hath seen it hath more Reason to believe it, than he that hath not'. His views were clearly set out in *The Sceptical Chymist*. While Boyle was at the height of his powers there occurred, in 1660, an event of most far-reaching scientific importance. This was the foundation in London of the Royal Society – fellowship of which is to-day one of the most highly prized of all honours in the field of science – formed specifically for the study of natural phenomena. This event has a double significance in relation to the advancement of chemistry; on the one hand it denotes the existence of a considerable body of men pledged to advance scientific knowledge, on the other it gave these men a very valuable means of exchanging views. This growth of interest in science was not peculiar to Britain; on the Continent, too, the study of chemistry was advancing rapidly.

It is not surprising that the process of combustion, which from time immemorial had provided man with such vital necessities as heat and light, and the means of cooking food and smelting metals, excited a lively curiosity. Not until the seventeenth century, however, was any plausible hypothesis put forward to explain this fundamentally important process. It was the great French chemist, Lavoisier, a victim of the French Revolution, who first realized what we now know to be the truth, namely that combustion is a process of combination with oxygen, a constituent of the air which Priestley had discovered in England.

These excursions into chemical theory were of great practical importance, for they were accompanied by a great increase in the number of substances known to chemists. In Sweden, for example, Scheele investigated hydrofluoric acid, chlorine, manganese, barium, oxalic and citric acids, and many other new substances. Priestley, whose discovery of oxygen we have mentioned, also studied other gases, developing special techniques for handling these elusive substances; among those he studied were hydrochloric acid, sulphur dioxide, nitric oxide, and ammonia. At Glasgow Joseph Black carried out important researches establishing the relationship between a number of important basic or alkaline substances, including limestone, lime, magnesium carbonate, magnesia, and soda.

All these great advances in practical and theoretical chemistry paved the way for one of the greatest of man's intellectual achievements, John Dalton's atomic theory, which for more than a century has profoundly influenced every kind of chemical thought. They could not have come at a more opportune moment. The last half of the eighteenth century, which saw these great chemical advances, was also the first half-century of the Industrial Revolution which has shaped our modern civilization. In the nineteenth century a very rapidly expanding industry suddenly turned chemistry from the hobby of a comparatively few scholars, and the source of supply of a comparatively small guild of craftsmen, into a science of immense practical importance in all branches of daily life.

CHAPTER TWO

Chemistry and the Industrial Revolution

THE Industrial Revolution, the beginning of which may be set at about 1760, had a profound effect on the simple chemical industry of the day. The rapid increase in the number of textile mills brought about a proportionate increase in the quantity of chemicals – such as soaps, mordants, alkalis, and acids – needed for processing the textiles at every stage of their progress from raw material to finished fabric. By the end of the eighteenth century great changes had been wrought.

The manufacture of sulphuric acid – to-day the most vital of all industrial chemicals – had been firmly established on a considerable scale in the first half of the century. In England, for example, there were sulphuric acid works in Richmond as early as 1736. It was made by dry-distilling iron sulphate; later by heating nitre and sulphur. Roebuck, who invented the important lead-chamber process, started a factory in Birmingham in 1746 and in 1750 this product began to be manufactured in Bradford. In 1790 Leblanc's discovery in France of a method of converting salt into soda by means of sulphuric acid brought about, in the course of a few years, a greatly increased demand for the latter substance. The establishment of the manufacture of alkali from salt is, however, of such great importance as to deserve a separate chapter and we shall revert to Leblanc's process and its effects later. By 1800 Britain was exporting 2,000 tons of sulphuric acid yearly.

One of the most important processes to which crude

textiles must be subjected is that of bleaching. In 1785 the French chemist Berthollet discovered the powerful bleaching action of chlorine, first made by Scheele in 1774. For practical purposes Berthollet recommended that chlorine used for bleaching should first be dissolved in potash solution, and he called the resulting liquid 'Eau de Javelle'. Berthollet described his discovery to James Watt, the engineer, who happened to be in Paris, and the latter brought news of it back to Glasgow, then a rapidly growing centre of the textile industry. The importance of the discovery was at once realized, for hitherto bleaching had been done chiefly by exposing material to the sun in bleach fields, though Roebuck had shown that linen could be bleached with sulphurous acid. The use of bleach fields required much space and, especially in northern latitudes, much time. An important improvement was soon made in Glasgow by Charles Tennant, who opened a chemical factory at St Rollox in 1799. He showed that chlorine could be absorbed by lime, yielding the very important product now known as bleaching powder. He began to manufacture bleaching powder in 1799, and in his first year produced fifty-two tons, selling at £140 per ton.

The increase in dyeing, a natural consequence of the development of the textile industry as a whole, led to a great expansion of what was already a very considerable industry. In the fifteenth century, for example, the discovery of alunite within papal territory caused Pope Pius II to establish the manufacture of alum and within a short time no less than 8,000 workers were being employed. In Britain its manufacture was established in Dorset, the Isle of Wight, and Yorkshire. Much later, in 1845, a completely new method of manufacturing alum was discovered by one of those fortunate accidents which, when they happen in the presence of somebody able to see their significance, sometimes achieve overnight what has been patiently but unsuccessfully sought over many years. In this case a Scottish chemist, Peter Spence, had been seeking for a means of making alum from coal-shale, but discouraged by his lack of success he was on the point of cutting his losses and turning his energy in other directions. By a mischance, however, a beaker of hot liquor which ought to have been thrown away was left in the laboratory overnight. As it cooled crystals of alum were formed and were noticed in the morning. By this lucky accident was founded a new industry for manufacturing alum by treating coal-shale with sulphuric acid.

Another important manufacture which was rapidly expanded to meet the almost insatiable demands of the new textile factories was that of soap-making. Although crude soap was known in the Sumerian era, and the Romans effected some improvements in soap-making, its manufacture does not seem to have been firmly established until about the end of the sixteenth century. The raw materials used were natural fats and oils and vegetable alkali made caustic by treatment with lime. The vegetable alkali used was derived from various sources, but chiefly from the burning of seaweed and of a plant of the goosefoot family. Even as late as 1834, when, as we shall see later, Leblanc's process for making soda from salt was firmly established, Britain imported no less than 12,000 tons of plant ash (barilla) from Spain. It is not known exactly when commercial soap-making was first established in Britain but at least one factory existed as early as 1703.

At the period we are now considering the only dyes available were of natural origin; not until after the middle of the nineteenth century were the first coal-tar dyes, now almost exclusively used, available in the dye-houses. Natural dyes in common use since ancient times were indigo and madder, giving respectively blue and red colours. Certain lichens, discovered in Asia Minor in the thirteenth century, were used as purple dyes. The scarlet dye cochineal was introduced into Europe from Mexico by the Spaniards in the sixteenth century; also from the New World came the black dye logwood. Other important natural dyes were cutch, turmeric, safflower (notable as a dye for government red tape), and litmus. Although all the dyes available were of plant or animal origin considerable use had to be made of manufactured chemicals. In consequence the dyehouses of Europe became important customers of the chemical industry. In the seventeenth century, for example, a Dutchman, Drebbel, showed that if cochineal is used with a tin salt and tartaric acid, a brilliant scarlet is obtained; this process was adopted for dyeing the uniforms of the British army, the first works for this purpose being erected at Bow in 1643. Dyeing with indigo was improved in the eighteenth century by adding ferrous sulphate and lime to the vat; in 1845 the zinc-lime vat was introduced. Chemists made detailed investigations of dyeing processes, effecting many improvements; among them may be mentioned Dufay, Chevreul, and Berthollet in France and Henry and Bancroft in England. Black dyeing was almost entirely a chemical process, the materials used being red iron oxide, gallic acid, and tannin.

The expansion of the chemical industry, although springing primarily from the immense growth in the production of textiles, was not confined to supplying the needs of this vigorous new industry. In many other directions, too, fresh ideas and greater prosperity were calling for new products to meet new needs. In what has been called 'the Age of Elegance', the manufacture of paint for decorative purposes naturally received a powerful impetus. By the end of the eighteenth century several newly discovered chemicals – such as Scheele's green, and yellow lead chromate – had been added to traditional pigments such as white and red lead, verdigris, and Naples yellow. Lead chromate was introduced by Kurtz, a Manchester chemical manufacturer, about 1800. He had an immediate success with it as, following the lead of Princess Charlotte, it became very fashionable to use yellow lead chromate paint for all kinds of carriages. A little later, in 1814, the French chemist Tessaert discovered a synthetic form of lapis lazuli in some furnace slag. In 1828 a method was worked out for repeating this fortuitous synthesis on an industrial scale. This brilliant blue pigment is made by heating china clay with sodium sulphate and charcoal, or in similar ways. Another valuable pigment dating from about the same time is Prussian Blue (iron ferrocyanide), yet another accidental discovery. By mixing Prussian Blue with yellow lead chromate a brilliant green – Brunswick Green – was made. Somewhat later, towards the end of the first half of the nineteenth century, an important new white pigment, zinc oxide, began to come into use. This has the advantage, compared with white lead, of not blackening in air containing traces of hydrogen sulphide; in the increasing industrialization of the nineteenth century, when the air in the manufacturing towns carried much sulphurous vapour, this was a valuable property. This increase in the range and quality of the pigments available was accompanied by an increasing skill in the blending of

CI–2

these with linseed oil, driers, and turpentine to form brilliant and durable paints. Thus the chemist was able to contribute notably to the colour of daily life – some compensation for the squalor and fumes which in the nineteenth century became the accepted accompaniment of most branches of industry. Later still, with the introduction of the first coal-tar dyes, the chemist was able to make a still greater contribution in alleviating the prevailing urban drabness.

While these slow but steady developments were taking place in the chemical industry, theoretical chemistry was going ahead very rapidly and in turn contributed in due course to new practical advances. As we have seen, Lavoisier and Priestley had at last revealed the true nature of the vitally important process of combustion. Black, at Glasgow, had elucidated the relationship between the various alkalis – soda, lime, magnesia, and so on. Cavendish, as well known for his personal eccentricity as for his contributions to chemistry, had discovered the composition of water, proving it to result from the combination of one volume of oxygen and two volumes of hydrogen. A very important change of outlook among chemists became apparent at this time; their work began to become quantitative rather than qualitative. Thus they were no longer content to know that if limestone was heated it lost carbon dioxide gas and left a residue of lime; they wanted to know what weight of carbon dioxide and what weight of lime was obtained from a given weight of limestone.

These quantitative investigations threw much light on the nature of chemical reactions and chemical compounds. It gradually became clear that substances did not combine together haphazardly, but in definite proportions. Water, for example, always consists of one part by weight of hydrogen to every eight parts of oxygen. When sodium and chlorine react to form ordinary salt twenty-three parts of sodium invariably combine with thirty-five of chlorine. These observations were of considerable intrinsic importance – for example, in developing new chemical manufacturing processes – but, as so often happens, it needed the touch of genius to see a common unity in all of them. In this case the genius was John Dalton, anticipated to some extent by William Higgins. It is by no means clear how Dalton came to formulate his atomic theory, which, in a somewhat modified form, has formed the foundations of chemical science for more than a century. Atomic theories were, it is true, well known, but only in a general and unsubstantiated form. Leucippus, Democritus, Epicurus, and, particularly, Lucretius had postulated a non-continuous state of matter in the days of classical Greece, well over two thousand years previously. Aristotle, whose views carried immense weight, had, however, ridiculed these early views and it was not until the seventeenth century that the atomic theory was revived, by Gassendi, a champion of Epicurus. Gassendi in turn was challenged by his own contemporary Descartes, who adopted a compromise. While not prepared to regard matter as consisting of atoms moving in completely empty space – a vacuum – he thought it might consist of discrete particles moving in some continuous aether. This dispute between two prominent scholars naturally attracted much attention and both sides gained many adherents, both in their own lifetimes and later. Boyle, 'the father of chemistry', was perfectly familiar with the arguments for and against an atomic theory. Its protagonists gained powerful support from Isaac Newton, who went so far as to interpret Boyle's law (which relates the volume of a gas to its pressure) mathematically on the assumption that gases consist of small, mutually repelling particles. Newton was, however, careful to state explicitly that while an atomic theory, interpreted mathematically, could explain the properties of gases the question of whether atoms really exist is one that 'philosophers may take occasion to discuss'.

What Dalton did was to crystallize the rather indecisive views of his predecessors and to formulate an atomic theory in precise and useful terms. Reflecting on some of the simple chemical reactions with which he was familiar, he supposed that the combinations involved, between say carbon and oxygen, took place in the simplest way imaginable, namely by the atoms pairing off. The atoms themselves he regarded as indestructible and incapable of being split into simpler bodies. Assuming this to be correct he could, from his knowledge of the relative combining weights of carbon and oxygen, at once calculate the relative weights of carbon and oxygen atoms. This was a tremendous step forward. All at once atoms had assumed inviduality and specific properties; no longer were they vague particles milling about in a sub-microscopic world of their own. New meaning was given to Boyle's conception of an element. Each element, according to Dalton, consisted of atoms of the same weight; different elements differed in the weights of the atoms of which they were composed. For the first half of the nineteenth century tables of atomic weights contained many uncertainties; these were not in fact resolved until the work of Cannizzaro, an Italian chemist, in 1858.

Although Dalton's work had far-reaching results, giving for the first time a logical explanation of the

nature of elements and why they combine in fixed proportions, it should not be assumed that these were
rapidly brought about. As late as 1851, for example,
the great Liebig's text-book of chemistry had room for
no more than a single paragraph on the atomic theory.
Nevertheless the sweep of the tide was irresistible, and
once uncertainty about the exact numerical values of
atomic weights was removed, the theory in its funda-
mentals gained, and has never lost, universal acceptance.

An important consequence of Dalton's theory was
that it made possible a kind of shorthand in which the
complex results of chemical reactions can be concisely
expressed. Every element was assigned a single letter,
or sometimes a pair of letters, to represent it; the letter
chosen was generally the first letter of the element's
name. Subscript numerals after each symbol indicated
the number of atoms of an element in a substance. For
example the formula H_2O for water means that each
'compound atom' or molecule of water contains two
atoms of hydrogen and one of oxygen. As_2O_3 for
arsenious oxide means that this substance contains two
atoms of arsenic to every three of oxygen. The theory
permitted also the writing of chemical equations to
express chemical changes both concisely and quanti-
tatively. This convenient shorthand has proved indispensable to the rapid development of chemistry.

Increasing interest in all branches of chemistry cul-
minated in the foundation of the Chemical Society, of
London. Until the early years of the nineteenth century
men of science had in the main had very broad interests, often undertaking impartially researches in what
we now regard as the separate realms of chemistry,
physics, and biology. Dalton, for example, wrote as
authoritatively on colour vision as on atoms. Gradu-

ally, however, as knowledge grew, scientists had to become more specialized and limit their interests to certain fields. Chemistry, formerly taught in the medical schools, emerged as a distinct science, though not·until 1845 was the first public school of chemistry opened in Britain; this was the Royal College of Chemistry. Chemists began to find the general scientific societies – such as the Royal Society and the Manchester Literary and Philosophical Society – unsuitable for discussing the details of their work. Accordingly some twenty-five leading chemists met in London in 1841 and resolved 'that it is expedient that a Chemical Society be formed'; Thomas Graham was elected the first president. So successful was it that it served as a model for the French Chemical Society (1857) and the German Chemical Society (1868). This important event marked the sure foundation of chemistry as a science in its own right; it marked also the beginning of a vast expansion in the chemical industry as a branch of applied science rather than as an empirical art.

CHAPTER THREE

The Growth of the Alkali Industry

T HE importance of alkali and the constantly increasing
need for it as the Industrial Revolution gained in
impetus have already been stressed. Very soon natural
alkali made by burning certain plants became inade-
quate to meet the need. It was in France – prevented by
war from importing soda freely – that the situation first
became acute and that active steps were taken to relieve
it. In 1775 the French Government, through the
Academy of Sciences, offered a prize of 100,000 francs,
at that time worth about £4,000, for the first person to
put forward a satisfactory industrial process for converting salt (sodium chloride) – readily available in
most parts of the world – into the much needed soda
(sodium carbonate). In offering this prize the French
Government was trying not so much to bring into
being a new process, as to give a powerful incentive to
a number of chemists who were already aiming at the
same target. As early as 1737, for example, Duhamel, a
Frenchman, had taken out a patent for carrying out
this very conversion, but although the process worked
well enough in the laboratory it was too difficult to
work economically on a large scale. In 1781 an English
chemist, Fordyce, was granted a patent for a somewhat
similar, but equally impracticable, process.

The coveted prize was finally awarded in 1790 to
Nicholas Leblanc. His process was worked for more
than a century; not until after 1900, after a hard fought
battle, was it decisively replaced by the modern

ammonia-soda process. Leblanc's process consisted in treating salt with sulphuric acid, thus obtaining sodium sulphate and, at the same time, clouds of hydrochloric acid gas. By roasting the sodium sulphate with limestone and coal a mixture known as 'black ash' was obtained; this consisted primarily of soda and calcium sulphide. The soda was extracted from the black ash with water, from which it was subsequently crystallized. Although many millions of tons of soda were made by this process before it became obsolete, the inventor himself had a tragic life. There is no record of the prize money ever having been paid to him. In 1791 the revolutionaries confiscated both his patent rights and the factory which had been built for him at St Denis by the Duke of Orleans, to whom he was surgeon. Eventually, under Napoleon, the works were returned to him, but without capital to work them his lot was no better than before. In 1806, overwhelmed by his misfortunes and completely destitute, he committed suicide.

Meanwhile interest in the salt-soda conversion had not by any means been confined to France; Fordyce's patent of 1781 has already been mentioned. Among those in Britain interested in chemical invention in the eighteenth century was the Earl of Dundonald, who impoverished himself as a result. In 1795, however, his prospects appeared good, for he took out a patent for the manufacture of caustic soda from salt and in partnership with John Losh set out to exploit it at their works at Walker-on-Tyne. Success appears to have been limited, however, for soon afterwards Dundonald withdrew from the partnership and John Losh then handed over control of the works to his brother William, who was evidently a man of enterprise. In

1802, after the conclusion of the Peace of Amiens, he was in Paris and became acquainted with the details of the Leblanc process, which he worked at Walker-on-Tyne after his return to Britain. It is thus to William Losh that we must give the credit for having introduced Leblanc's process into Britain. As this was the most important industrial chemical process of the nineteenth century, influencing the development of the industry as a whole, it deserves discussion in some detail.

Despite its importance the process was only very slowly adopted by other manufacturers. In 1814 Lutwyche and Hill in Liverpool, suppliers of alkali to a number of soapmakers in that city, worked the Leblanc process on a small scale. The first to work it on a large scale was a remarkable man named James Muspratt, born in Dublin in 1793. At the age of 14 he made his first acquaintance with chemistry by being apprenticed to a druggist in Dublin. This life, however, proved too dull for him and he applied for, but failed to get, a cavalry commission in order to fight in the Peninsular War; nevertheless he followed the army through much of the campaign. After many adventures he found his way back to Dublin and there set up as a small chemical manufacturer, with a partner named Abbott. This partnership was dissolved in 1822 and Muspratt went to try his luck in Liverpool. In passing, it may be remarked that the whole story of nineteenth-century chemical industry presents a bewildering picture of changing partnerships; many of the leaders of the industry were men of strong and colourful personality, not readily subduing their own impulses when working with others. Such clashes of personality were no doubt beneficial rather than otherwise to the young industry, for they resulted in developments quickly occurring in more directions than might otherwise have been the case. On his arrival in Liverpool Muspratt acquired an old glassworks and there made potassium prussiate. When he had acquired a little capital he embarked, in 1823, upon soda manufacture by the Leblanc process. Despite the excellence of his product he found difficulty in selling it to manufacturers – primarily soap-makers – who had long been accustomed to the imported barilla and similar vegetable products.

Initial difficulties overcome, Muspratt began to manufacture Leblanc soda on a large scale, both in Liverpool and at St Helens. Unfortunately, however, this very extension of his activities increased his difficulties, for the vast quantities of noxious hydrochloric acid evolved during the treatment of salt with sulphuric acid caused, very naturally, much annoyance to local landowners, who sought redress in the courts. By 1838 public resentment had reached such a pitch that, after many lesser battles, Muspratt was indicted before a special jury at Liverpool Assizes. Muspratt's counsel could not deny the reality of the nuisance, though some half-hearted attempt was made to blame the desolation of the neighbourhood on 'salt brought from the sea by violent gales' and sought to justify his client by pointing out – with much reason but little legal force – the immense national value of the alkali industry. This, and other litigation, had considerable influence on the early development of the chemical industry, particularly upon the siting of the works.

It must not be thought that Muspratt and other early alkali manufacturers were indifferent to the fate of their unfortunate neighbours; the difficulty was that there was then no known method of absorbing the noxious vapour or of dissipating it harmlessly in the atmo-

sphere. The choice lay between annoying the neighbours and closing the works, and the manufacturers
would have been more than human had they accepted
the second of these alternatives. Muspratt put his faith
in very tall chimneys – one was almost three hundred
feet high – for dispersal of the fumes, but under certain climatic conditions the vapours discharged were
carried down upon the surrounding countryside instead of up into the atmosphere.

The first man to devise an effective means of absorbing the offensive hydrochloric acid was William Gossage, born in 1799. In 1830 Gossage began to manufacture soda in Worcestershire; he was very soon made
aware of the unpopularity of the works with his neighbours and sought a remedy. Near the factory stood a
derelict stone windmill and Gossage hit upon the idea
of knocking the floors out of this and filling the interior
with brushwood. The waste gas passed upward through
this tower, against a descending stream of water; the
latter effectively washed out the acid. Gossage patented
his invention in 1836; in the first Alkali Act of 1863
manufacturers were required to absorb at least 95 per
cent of the hydrochloric acid in their waste gases.

Like other manufacturers of his day Gossage seems
to have had a restless disposition. He moved to Birmingham in 1841 and manufactured white lead. Three
years later he was copper smelting in South Wales. In
1848 he came back for two years to Worcestershire and
then moved to Widnes, then just metamorphosing
from a quiet country village to a main centre of chemical industry. In Widnes he worked the Leblanc soda
process and also extracted sulphur from copper ores.
By 1854, however, he was doing something quite
different. The rising price of tallow, a consequence of

the Crimean War, led him to set up as a soap-maker and his mottled soap soon became world famous. His soap works remained in existence until 1932.

Another colourful pioneer of the alkali industry was Josias Gamble. He differed from most of his contemporary manufacturing colleagues in having had a thorough education in chemistry, studying under the famous Dr Cleghorn at Glasgow. Like Muspratt, he first set up as a chemical manufacturer in Dublin, making bleaching-powder, sulphuric acid, alum, and Glauber's salt. Later he went into partnership with Muspratt at the St Helens works, but this lasted only two years. Gamble retained the St Helens works and in 1835 purchased a nearby lead-chamber works for manufacturing sulphuric acid.

It should, perhaps, be remarked that at this time sulphuric acid, an essential material for the Leblanc process, was for the most part manufactured by the alkali manufacturers within their own works by the lead-chamber process invented by Roebuck.

Because of its subsequent importance in the development of the ammonia-soda process, which eventually succeeded Leblanc's, mention must be made of another series of relationships which existed in the early days of the alkali industry. About 1830 there arrived in Britain a German chemist, Andrew Kurtz, who had already seen fifty years of adventurous life. On his arrival in Britain he set up as a small chemical manufacturer at Liverpool, making pigments. By some means not now clearly understood he established a relationship with a man named Darcy, who had a soda factory at St Helens. Despite Kurtz's bitter denials Darcy insisted on claiming him as a partner in this business, which traded as Darcy and Kurtz. Whatever the rights or wrongs of the

matter may have been, Kurtz was obliged to take over
the St Helens works when Darcy became bankrupt in
1842. Saddled with this business, Kurtz sent his son to
Paris to study chemistry under Gay-Lussac; there the
son met another English student, John Hutchinson.
This friendship led eventually to Hutchinson's employment in Kurtz's factory at St Helens but, Kurtz dying
shortly after his appointment, Hutchinson set up, in
partnership with a lime-burner named Earle, at nearby
Widnes. There he acquired the services of a number of
men who were later to prove of great distinction in the
chemical industry. His first manager was Henry Deacon
who, during later partnerships with William Pilking-
ton, of the glass-making family, and Holbrook Gaskell,
unsuccessfully pursued the elusive ammonia-soda process. Paradoxically enough, however, Deacon did in-
vent an important modification of the Leblanc process which gave it a new lease of life in face of competition from the ammonia-soda process when the latter
was eventually perfected by the Solvay brothers in
Belgium. The Solvay process was introduced into
Britain, under licence, by John Brunner and Ludwig
Mond, both of whom were at one time associated with
Hutchinson, one as employee and the other as partner.

The story of the long battle between the Leblanc and
the Solvay processes for making soda must be told in
detail, for it had a profound influence on the orientation
of the industry as a whole. It is indeed not too much to
say that it has affected not merely the history of the
chemical industry but the history of the world. To
understand why the Solvay process finally displaced
the Leblanc we must study the chemistry of the latter
in greater detail than has so far been necessary. In the
Leblanc process, it will be remembered, the soda was

finally obtained by extracting the black ash with water. This left behind, however, a considerable quantity of alkali waste; in fact for every ton of soda produced no less than two tons of alkali waste was formed. Not only was this waste a most noxious material, presenting considerable problems in its disposal, but it was also a serious source of loss, for it contained almost all the expensive sulphur originally present in the sulphuric acid used for making the salt-cake in the first stage of the process. Despite many attempts it proved impossible to recover this sulphur economically for further use. Besides being inefficient in its use of materials the Leblanc process also entailed the constant risk of litigation with offended neighbours.

It is, therefore, not altogether surprising that an alternative to the Leblanc process was sought at an early date; the search began indeed even before the Leblanc process was firmly established. As long ago as 1810 the French scientist Fresnel proposed the series of chemical reactions which form, in its essentials, the modern Solvay process. Technical difficulties, however, proved immense and more than half a century passed before theory could be translated into practice.

The basis of this very elegant process is to treat salt with ammonia and carbon dioxide, thus forming ammonium chloride and sodium bicarbonate. If the latter is heated it is converted into soda (sodium carbonate) with loss of carbon dioxide, which goes back to be used again in the first stage of the process. The ammonium chloride can be heated with lime, yielding calcium chloride and ammonia; the latter too can be circulated back for re-use, theoretically without loss though in fact small additions have to be made from time to time. Thus the only product not utilized is the calcium chloride, which is not only quite innocuous but also does not lock up any chemicals of particular value.

The story of the pursuit of this attractive process is for the most part one of broken hearts and broken fortunes. In 1836 an attempt, financially disastrous, was made to work it in Scotland. James Muspratt lost what was then the very considerable sum of £8,000 in another unsuccessful attempt. In Vienna, Leeds, and Paris more money was lost. Not until 1872 was success achieved, by Alexander and Alfred Solvay in Belgium. Once achieved, success was spectacular; within a few years the Solvay patents were being worked throughout the world – in Russia, Germany, Austria, Britain, the United States, and elsewhere. Before the onslaught the Leblanc process gradually collapsed. In 1902 the world's factories produced 1,800,000 tons of soda; of this no less than 1,650,000 tons were made by the Solvay process, only thirty years after the granting of the first comprehensive patent. Not until after the end of the First World War, however, was the Leblanc process finally extinguished.

It must not be thought, however, that the surrender was a tame one. Throughout the period in which the Leblanc soda process reigned supreme the manufacturers constantly sought to improve it, in order both to meet their competitors and to meet the challenge of an alternative process if and when it arose. Very naturally much of the research for an improvement centred about the recovery of the valuable sulphur from the alkali waste. Ludwig Mond, who later pioneered the Solvay process in Britain, half solved the problem in that, by a complicated series of operations, a method was found for recovering half the sulphur. John Hutchinson gave Mond's method a trial in his Widnes works and Frederic

Muspratt and others worked it under licence. Mond's patent was taken out in 1862, but mention ought also to be made of an earlier one by William Gossage, taken out in 1857.

Although Gossage's method was never successfully worked, Alexander Chance, the chemist who finally solved the problem of sulphur recovery from alkali waste, paid a handsome tribute to it. He expressed the opinion that Gossage's process failed only because at the time it was taken out chemical pumps and machinery were of too primitive a nature to permit its success; there was nothing inherently wrong in the process. This, it may be remarked, is a common happening in the chemical industry. Many an excellent chemical process has had to be shelved through lack of practical means of exploiting it.

In 1870 Weldon – a versatile chemist whose activities included the publication of *Weldon's Ladies' Journal* – patented a process which made a great improvement in the conversion of surplus hydrochloric acid into the readily marketable bleaching powder. At that time the acid was converted into chlorine by oxidizing it with manganese dioxide, but this had the disadvantage of leaving a residue which locked up expensive manganese just as the alkali waste locked up sulphur. Weldon's process made possible the recovery of the manganese so that it could be used again. This had the immediate effect of increasing the world's output of bleaching powder and of reducing its cost. Dumas, the great French chemist, said of Weldon's discovery: 'By Weldon's process every sheet of paper and every yard of calico has been cheapened throughout the world'; such can be the influence of a new chemical process. At about the same time Henry Deacon patented a process for oxidizing hydrochloric acid to chlorine without the use of manganese at all.

Despite these improvements the Leblanc process soon felt the weight of the Solvay patents. Accordingly, Alexander Chance's discovery in 1882 of an economic method of recovering the whole of the sulphur from the alkali waste was most timely and enabled the Leblanc process to struggle on for a few more years.

Although the final result of the struggle must even then have seemed certain, those working the Leblanc process used every possible device to compete with the Solvay. In Britain the struggle eventually resolved itself into one between the United Alkali Company, and Brunner, Mond and Co. The latter, the British licensees of Solvay, starting precariously on very limited capital and in face of an almost overwhelming series of practical difficulties, had begun to operate the ammonia-soda process at Winnington in 1873. Both partners reaped a rich reward for their perseverance and courage. The wealth which came to them was used wisely. Mond settled in London and became a patron of the arts and science. He formed a magnificent collection of early Italian paintings, which he bequeathed to the nation, and presented a magnificently equipped chemical laboratory to the Royal Institution. His name is remembered also for his important processes for the extractions of nickel from its ores and for the production of producer gas, both widely used in the modern chemical industry. Brunner settled in Cheshire, and represented the Northwich Division in Parliament for many years. He was a generous supporter of Liverpool University, in which he endowed chairs of Egyptology, Physical Chemistry, and Economics.

The United Alkali Company was formed in 1890 and

was an amalgamation of some forty-five firms, practising the Leblanc process, which hoped to achieve more efficient working by pooling their resources, closing some factories, and making drastic economies. In this amalgamation many famous names – such as Gaskell and Deacon, Muspratt, and Hutchinson – were lost. Brunner, Mond and Co. in turn extended their interests by from time to time buying up one or other of the remaining independent soda manufacturers. In 1911, for example, they acquired the Castner-Kellner works. In 1917 they acquired Chance and Hunt, then almost the only important alkali manufacturers remaining outside the two great combines. In 1926, the long and unproductive rivalry ceased. Brunner, Mond and Co. and the United Alkali Co. merged with two other great chemical industrial firms to form Imperial Chemical Industries Ltd; this very significant merger will be discussed later.

The final product of both the Leblanc and the Solvay process is soda ash, which is familiar in a hydrated form as washing soda. For many purposes, however, such as soap-making, paper-making, oil purification, and treating textiles, a more powerful alkali – caustic soda – is needed. The traditional way of making this is by treating soda ash with lime. To-day, however, much caustic soda is made by an electrical process, for which we are indebted to a brilliant American chemist, Hamilton Young Castner, who landed at Liverpool in 1886. In his own country he had perfected a process for making metallic sodium by heating caustic soda with iron and carbon. Failing to find financial support at home he came to England and made an agreement for the working of his process with a Birmingham firm which was using large quantities of sodium for making the then comparatively little known metal aluminium. It was paying 14*s*. per pound for sodium; after adopting Castner's process it paid 1*s*. For the first time aluminium became available at a reasonable price. Castner was, however, extremely unlucky, for almost at once Hall in America and Héroult in France independently perfected an electrolytic process for making aluminium, with which the British process could not compete. Accordingly Castner looked for other outlets for his metallic sodium. He invented a process for converting it into sodium cyanide by reaction with charcoal and ammonia, and made arrangements with the Cassel Gold Extracting Company of Glasgow to develop this commercially. A great potential market existed in the various gold-fields which were using potassium cyanide for recovering gold from crushed ores; unexpected difficulty was experienced in persuading the customers to accept the sodium salt in place of the potassium, although it was in fact more efficient. Castner eventually overcame the difficulty by the brilliant expedient of labelling his sodium cyanide '130 per cent potassium cyanide'. In American the cyanide process was worked at Niagara Falls; factories were also established in Germany. To-day the manufacture of cyanides, primarily for gold extraction and for electroplating, is a very important part of the chemical industry.

The manufacture of cyanides proved so successful that Castner set out to improve upon his original method of manufacturing metallic sodium. For this he turned to a discovery, made originally by Davy in the early nineteenth century, that if an electric current is passed through molten caustic soda metallic sodium is liberated. Although he developed this process into one capable of being worked industrially he experienced great difficulties owing to the impure nature of the best caustic soda available to him. Accordingly, he set out to find a way of making purer caustic soda. His ultimate solution was an electrolytic cell, containing mercury, in which an electric current was passed through brine. This was not an entirely new principle, but was certainly its first successful exposition. On taking out world patents for his invention Castner encountered difficulty in Germany, where an Austrian chemist, Carl Kellner, had just filed somewhat similar patents. To avoid litigation – though it now appears that Kellner would have had great difficulty in substantiating his claims – a compromise was reached. This type of electrolytic cell, first operated in 1886 and now in use throughout the world, is therefore usually known as the Castner-Kellner cell. Besides producing caustic soda of a purity far higher than any previously available in the trade, the cells produce as by-products chlorine and hydrogen. The modern disposal of these will be discussed in a later chapter.

A mild form of alkali, of great industrial significance, is lime, which is made by roasting limestone or chalk (calcium carbonate) in kilns. In many parts of the world – for example in Somerset and Derbyshire – whole ranges of hills consist of almost pure calcium carbonate. Huge quantities of limestone are used for making cement, following the invention of Portland cement by Aspdin in 1827. Lime is used for making bleaching powder, for fertilizing sour soils, for causticizing soda, for converting raw hide into leather, and – outstandingly important – as a means of forming the slag in steel making. The Bessemer process requires half a ton of limestone for every ton of iron converted into steel.

CHAPTER FOUR

Sulphuric Acid – The Barometer of Industry

ALTHOUGH sulphuric acid is a chemical rarely encountered as such in daily life it is one of the most important of all industrial chemicals, for an immense number of important processes depend upon its use at one stage or another. In the previous chapters we saw its vital importance in the manufacture of soda by the Leblanc process. The acid finds many other uses – for example in making dyes and drugs, leather processing, oil refining, paint manufacture, iron pickling, fertilizer manufacture, and the extraction of metals. Lord Beaconsfield was well aware of the immense industrial importance of sulphuric acid when he wrote: 'There is no better barometer to show the state of an industrial nation than the figure representing the consumption of sulphuric acid per head of population'. The choice of the word consumption was not perhaps the happiest possible when speaking of so corrosive a fluid as sulphuric acid, but it may nevertheless be remarked in passing that the average Englishman to-day consumes roughly $2\frac{1}{2}$ lb of sulphuric acid each year. Sulphuric acid, although far less familiar to us than the common washing soda which formed the subject of the previous chapter, is no less deserving of a separate description.

Some reference has already been made to early ventures in sulphuric acid manufacture. Ward set up a factory at Richmond, in Surrey, in 1736; Rawson manufactured at Bradford in 1750; and Kingscote and Walker began to make the acid at Battersea in 1772. By far the

most outstanding figure in these early days was, however, John Roebuck, who introduced what is called the
'lead-chamber process' in 1746; this, in a modified
form, is still of great importance.

Ward's method consisted in burning a mixture of
brimstone (sulphur) and nitre in the necks of a battery
of 66-gallon glass vessels containing a small amount of
water. This was a modification of a process familiar to
Glauber as early as 1648. After several combustions the
water in the flasks was converted, owing to absorption
of the oxide of sulphur formed by the burning of the
brimstone in the presence of the nitre, into dilute sulphuric acid; this was then concentrated by distillation.
From the industrial point of view, however, this was not
a very satisfactory process; the glass vessels were cum-
bersome and fragile and even after several combustions
each yielded no more than four gallons of dilute acid.
Nevertheless, it is noteworthy as the first process
worked on an industrial scale and it reduced the price
of the acid from £2 to 2s. per pound. Ward did not
patent his process until 1749, when it had already been
made obsolescent by Roebuck's 'leaden houses'.

Roebuck utilized Glauber's observation that lead
was resistant to sulphuric acid; moreover it was far
more robust than glass and was easily worked, especially after the introduction of the hydrogen blow-pipe in
1838. He appears to have developed his interest in sulphuric acid through a chemical consulting business in
Birmingham, which he had taken up after abandoning
medical studies; many of his clients required assays of
the sweepings from the jewellers' workshops which
were springing up in large numbers in Birmingham.
His first lead chambers were quite small – of no more
than 200 cu. ft capacity – and consisted of lead sheeting

supported by an external wooden frame. In the bottom
of the chambers was about 4 in. of water; a mixture of
sulphur and nitre could be burnt in them in a dish supported on a lead pedestal. The chambers were arranged
in batteries of ten, working day and night; at the end
of a month of successive combustions the dilute acid
was removed from the bottom of the chambers and
concentrated in lead vessels.

Roebuck soon established a works at Prestonpans,
where he hoped to be able to preserve the secret of his
process, which he had not then attempted to patent,
better than in Birmingham. In this he was unfortunate,
for a faithless employee left him and set up lead chambers at Bridgnorth, on the Severn. Another man also
obtained the secret by bribing Roebuck's employees,
and set up a works at Govan. Rival works also sprang
up at Dowles and elsewhere. In 1766 a lead-chamber
works was in production in France, and the Stirlings of
Glasgow began to use the process to make acid for
bleaching purposes. Thus the secret became generally
known and when Roebuck made a belated patent
application his rivals were able to send witnesses to the
Scottish courts to show that the lead-chamber process
was already in general use, and Roebuck's application
was rejected. Free of patent restrictions the lead-chamber process was extensively worked in Britain. By 1805
a works using 305 chambers was in action at Burntis-
land. Roebuck's own works, which he sold about 1785,
had more than 100 chambers by the end of the century.
A considerable export market for sulphuric acid was
established.

Improvements were slowly made. In 1793 the French
chemists Clément and Desormes showed that air could
replace nitre for the combustion of the sulphur, though

small quantities of nitre were found to be essential for the necessary chemical reactions to take place. In 1827 Gay-Lussac devised a tower for absorbing the nitrous fumes, both expensive and noxious, which had formerly been discharged into the air; this tower did not, however, come into general use until 1859, when Glover invented a denitrifying tower, the use of which much increased the value of Gay-Lussac's. The size of the chambers was increased; in 1860, for example, Muspratt built a single chamber which was 140 ft long, 24 ft wide, and 20 ft high. Another very important innovation, and another triumph for British chemistry, was the introduction of platinum stills for concentrating the dilute acid.

Yet another important development, dating from 1818, was the introduction of pyrites in the place of sulphur; this innovation was originally made by Hill, of Deptford. This was followed by various inventions aimed at the more effective burning of pyrites, this being done in separate burners instead of inside the chambers. Improved types of sulphur burners were also designed. Other improvements related to methods of concentrating the acid; in some, such as those designed by Kessler and Gaillard, hot furnace gases, formerly wasted, were used to evaporate excess water.

These various improvements and modifications, which involved, however, no fundamental change from Roebuck's method, were very largely British in origin, with the result that the lead-chamber process became known as the 'English process'. In Britain it held the field up to the outbreak of the First World War. On the Continent, however, a rival process began to be exploited in the last quarter of the nineteenth century; it is, however, noteworthy that this process too was British in conception.

The new process was called the contact process and was first described in a patent lodged in 1831 by a Bristol vinegar manufacturer named Peregrine Phillips. In view of the immense importance of this process today it is a matter for regret that scarcely anything is known of Phillips' life. To understand the nature of his process we must go a little more deeply into the chemistry of the lead-chamber process. The latter may be regarded as taking place in three steps. In the first, sulphur, either as brimstone or pyrites, is burnt to form sulphur dioxide gas. In the next stage the sulphur dioxide is still further oxidized, through the action of the nitrous fumes, to sulphur trioxide, which in its pure state is a white solid. In fact, however, the white solid does not actually appear, for the third stage – combination of sulphur trioxide and water to form sulphuric acid – occurs immediately. Phillips discovered that the conversion of sulphur dioxide to sulphur trioxide could be effected without the intervention of nitrous fumes simply by passing a mixture of air and sulphur dioxide gas over finely divided platinum contained in a heated tube. The platinum was not consumed in the process but acted as what is called a catalyst – a substance which by its mere presence facilitates the chemical combination of other substances. Furthermore this process makes possible the immediate production of highly concentrated sulphuric acid, known as oleum, without the need to evaporate dilute acid, a process expensive in terms of both time and fuel. There is no evidence that Phillips ever worked his process industrially, but as early as 1847 a Belgian chemist named Schneider, following experiments of Magnus and Döbereiner in

Germany, set up a small demonstration plant. Soon afterwards Wöhler and Mahla showed that iron oxide, and certain other relatively cheap oxides, could be used as a catalyst in place of the expensive platinum specified by Phillips. The first industrially successful plant for making acid by the contact process was set up in Russia.

Nevertheless, the contact process still offered no serious challenge to the lead-chamber method. The latter was effective and economical and it yielded an acid satisfactory for the industrial needs of the day, most particularly for the Leblanc soda process and the making of fertilizers. There was no great reason, therefore, for manufacturers to spend great sums on research and equipment to launch a new process which was as yet almost untried and offered no obvious advantages. In the 1870's, however, circumstances changed. In Germany, inspired by the discovery of the first synthetic dyestuff in Britain in 1856, a great new organic chemical industry was developing. For this industry the highly concentrated form of sulphuric acid known as oleum presented a great advantage over the strongest acid yielded by the lead-chamber process; in consequence the time was ripe for the exploitation of the contact process.

One difficulty encountered in the contact process was that after a time the catalyst became ineffective – it was 'poisoned'. In 1870 a German chemist, Rudolph Messel, at that time employed by a firm of chemical manufacturers in Stratford, discovered that the poisoning of the catalyst was due to the presence of impurities in the gases; if the latter were carefully purified beforehand the chemical reaction would go on indefinitely. In 1875 Messel, with one of the partners in his firm, lodged a patent to protect this discovery; a plant to manufacture oleum was set up at Silvertown, eventually producing 1,000 tons weekly. It is interesting to note, as a measure of the importance of oleum, that originally they obtained their mixture of pure sulphur dioxide and oxygen by decomposing lead-chamber sulphuric acid. Later, however, they used sulphur dioxide obtained by burning sulphur. To economize in the very expensive platinum it was used in a very finely divided form spread on asbestos fibres. In 1878 Messel succeeded Squire as director of the firm, which prospered exceedingly under his direction. Just as Brunner and Mond – the pioneers of the new ammonia-soda process in Britain – used a great part of their wealth for the benefit of their fellow men, so Messel, the British pioneer of the no less important contact process, devoted the bulk of his very considerable fortune to useful ends; at his death in 1920 he left large sums to the Royal Society and to the Society of Chemical Industry.

Meanwhile, in Germany, the contact process was being even more fully exploited to meet the needs of the rapidly expanding dyestuffs industry. The most important centre of the industry was at Hoechst-am-Rhein. At Ludwigshafen, too, intensive research to improve the process was undertaken in the laboratories of the Badische Anilin und Soda Fabrik, a firm which then required large quantities of oleum for the manufacture of synthetic indigo. An important discovery made there was that the heat released when the sulphur dioxide combined with oxygen on the catalyst was sufficient, if carefully conserved, to maintain the temperature of the gases at about 400–500° C., the temperature at which the highest yield of acid is obtained. In this way a valuable saving in fuel was obtained, with a consequent reduction in working costs.

CHAPTER FIVE

Natural and Synthetic Carbon Compounds

Up to the present we have been considering almost exclusively substances of mineral origin – such as salt, soda, sulphuric acid, copper sulphate, alum, and so on – which are all derived, directly or indirectly, from the inanimate material of the earth's crust. These are known to the chemist as inorganic substances. There is, however, another great class of substances to which we have so far only occasionally referred because it is only comparatively recently that they have, in the chemical sense, become of outstanding importance. These other substances are the organic chemicals, originally so called because they were derived from living organisms. This class includes the natural textile fibres such as wool and cotton; certain natural drugs such as quinine; acids of vegetable origin such as tartaric and citric acids; natural plastics such as rubber; fats and oils; and so on. Despite their great variety, which even the above very incomplete list illustrates, organic chemicals have one important characteristic – they contain the element carbon. To-day we know, thanks to the ingenuity of chemists, a vast number of substances containing carbon – dyes, plastics, drugs, insecticides, weed-killers, photographic developers, and many others – which are not known in nature at all; nevertheless they are called, for the sake of convenience, organic chemicals.

Carbon is quite the most remarkable of all the ninety-two natural elements. To a degree quite unknown in all other elements, it has the power of combining freely with itself; atoms of carbon can form long chains, branched or straight, as well as large and small rings or combinations of rings. It can join with a great variety of other elements too, such as nitrogen, hydrogen, oxygen, sulphur, chlorine, bromine, and iodine. Thanks to this unique versatility carbon can form hundreds of thousands of different compounds – perhaps millions – whereas the compounds of any other single element are, comparatively speaking, numbered in scores. So large is the number of carbon compounds now known, both natural and artificial, that their study needs the attention of specialists. Methods have been worked out for discovering the exact pattern of the carbon and other atoms in all but the most complex organic substances and a simple and logical classification may be made in terms of these patterns. For practical purposes the vast number of organic substances can be classified within a comparatively small number of different classes.

In speaking of rings of carbon atoms, of synthetic organic dyestuffs, drugs, and so on, however, we have been running ahead of the period with which we are now dealing. In the earliest days of the chemical industry organic chemicals were exactly what their name implied – chemicals derived from living organisms, either plants or animals; synthetic organic chemicals are essentially a creation of the last hundred years.

Certain organic chemicals are of very ancient origin. Many natural drugs have been in use since the earliest known times. Opium, for example, is clearly referred to in Homer's Odyssey. This drug is spoken of as coming from Egypt, 'whose rich earth herbs of medicine do adorn in great abundance'. The Assyrians used opium suppositories to treat abdominal pains. The common hemlock, containing a poison named coniine, was well known to the Greeks, who used it for state executions; Socrates died in this way in 402 B.C. The root of the mandrake was used in ancient times as a pain-killer, especially for relieving toothache and for minor surgical operations. In the course of centuries other vegetable drugs gradually came into use. From the New World, for example, the Jesuits brought quinine; cocaine, too, was first introduced into Europe from South America. Digitalis, the foxglove drug, long a traditional remedy for dropsy, achieved medical recognition in the eighteenth century thanks to the Birmingham physician William Withering.

By the end of the eighteenth century, therefore, the preparation of a considerable range of organic drugs, supplementing the inorganic ones which we have already mentioned, was an important, though small, industry. It was for the most part in the hands of small family units. Many of these, expanding their interests to keep pace with the advance of medical science, are still in existence and internationally known.

The investigation of anaesthesia in the second quarter of the nineteenth century, culminating in James Simpson's discovery, by personal experience, of the anaesthetizing properties of chloroform, widened the scope of the pharmaceutical chemical industry. At about the same time increasing knowledge of the value of antiseptics, though these were not clearly understood until the famous bacteriological experiments of Pasteur, still further broadened the field of the chemical industrialist. The value of bleaching powder (chloride of lime) for preventing the spread of infection was recognized as early as 1830 and in 1832 the British

Board of Trade experimented with the fumigation of ships with chlorine. In France, too, chlorine in various forms found increasing use for sanitary purposes.

Lister, anticipating Pasteur's discovery of the bacterial nature of infection, introduced carbolic acid in 1865 as a surgical disinfectant, with spectacular success. Soon chemical relations of carbolic acid – called the phenols – came into use. For these substances manufacturers turned to coal-tar, the black viscous material obtained as a by-product of the then comparatively new gas industry. For the latter we are indebted to William Murdoch, the pioneer of gas-lighting. In later years, following Perkin's discovery of the first coal-tar dye in 1856, black and apparently useless coal-tar was to prove to the organic chemist an almost unimaginably rich treasure-house.

At about the same time the birth of chemistry as an independent science, marked by the foundation of the Chemical Society in 1841, led to an increased demand for highly purified organic chemicals for analytical and similar purposes. Other demands for highly purified chemicals, organic and inorganic, resulted from the increasing use of photography following the patenting of Fox-Talbot's calotype process in 1841. This followed the experiments of earlier investigators of colour-sensitive chemicals, notably the Frenchman Daguerre.

By 1856, therefore, a considerable organic chemical industry had been established. At the same time laboratory research on organic substances was being intensively pursued by chemical research workers. William Perkin's epoch-making discovery of the first useful synthetic organic dyestuff in that year therefore occurred at a time when the chemical industry was not without experience in the methods necessary to translate this accidental laboratory discovery into an important manufacturing process.

At the time of this discovery Perkin was only 18 years of age, a student of chemistry under the great German chemist Hofmann, who had been invited to London to be the director of the newly founded Royal College of Chemistry. By what we now know to be a quite impossible method Perkin was trying, during a holiday spent in his father's house, to make quinine from aniline. Nevertheless, from the black and unpromising-looking mess which resulted from one of his experiments he isolated a reddish powder which proved to be a powerful mauve dye. Young though he was Perkin at once realized the commercial possibilities of this discovery and he submitted a sample of it to Pullars of Perth, a famous dyeing firm. Receiving a favourable report he set up a factory at Greenford Green, near London, to manufacture mauve. Curiously enough, mauve was not immediately popular in Britain, but fortunately became extremely fashionable in France, whence it was brought back to fashionable circles in London.

Perkin's exploitation of his discovery was not, however, made without difficulty. To make mauve economically he required considerable quantities of aniline at a moderate price. This valuable oily liquid was first obtained in 1826 by Unverdorben, who made it by heating natural indigo until it decomposed. Soon afterwards it was found that aniline could be made from nitro-benzene, itself prepared by treating benzene with a mixture of strong sulphuric and nitric acids; it is interesting to note how almost inevitably we come back, at some point or another, to sulphuric acid. Benzene had been discovered by Michael Faraday in

1825, as a constituent of gas made from whale oil; it was later found to be present in considerable quantities in coal-tar. It was, therefore, to the tar-distilling industry, then of quite limited importance, that Perkin turned for benzene as the ultimate source of his new dye.

Chemists and industrialists everywhere soon saw the immense prospects which Perkin had opened up. By replacing aniline with various other chemically related substances, such as toluidine and quinoline, a whole range of new colours was obtained. Quinoline Blue, Imperial Violet, Magenta, and a number of others were discovered within a few years and manufactured in factories which sprang up all over Europe – at Lyons, Ludwigshafen, Manchester, and elsewhere. Before the possibilities of this class of dyes were exhausted, Heinrich Griess, who also had been studying under Hofmann in London, discovered quite a new class, known chemically as azo dyes. Typical of these was Bismarck Brown. Griess, a veritable caricature of the eccentric scientist, was a man of great brilliance and made many notable discoveries in the dyestuffs field.

In 1869 came a sensational new discovery. Perkin in England and Caro in Germany independently discovered how to synthesize alizarin, the natural dye of the madder plant and one used extensively since the very dawn of history. This was a dramatic story. Perkin was extraordinarily unlucky, for when he went to the Patent Office in London to lodge his patent he found he had been anticipated, by one day only, by Caro. This was a heavy blow, for it meant that the extremely profitable British market for synthetic alizarin was wide open to Germany. The effects on the madder-growing industry were immediate and catastrophic;

importation of madder into Britain, chiefly from France, slumped from several thousand tons a year to almost nothing. The industry for producing the natural dye was killed almost overnight. No less disastrous, this time to Indian growers, was the synthesis of indigo in 1880.

Caro's discovery of synthetic alizarin is of such interest and importance that it deserves further notice. A calico-printer by training, he came to Manchester in 1856 but, as a result of the virtual collapse of the synthetic dyestuffs industry in Britain, returned to Germany in 1867 to carry out research at Heidelberg. While there he became consultant to the Badische Anilin und Soda Fabrik, eventually becoming head of the research laboratory at Mannheim-Ludwigshafen. It was there that the synthesis of alizarin was perfected; the process eventually used owes much, like Perkin's original discovery, to a fortunate accident. Caro was trying to make a new dye by heating a substance called anthraquinone with a mixture of sulphuric and oxalic acids. In the middle of his experiment he was called away and forgot to remove the flame from beneath the vessel. On his return the mixture had boiled dry and the laboratory was filled with choking fumes. Round the edge of the charred mess in the dish, however, Caro noted a pink line which eventually proved to be an aluminium derivative of alizarin. Working from this chance discovery he perfected a workable process for the synthesis of alizarin.

So far as Britain is concerned, however, the story of the next fifty years was a sorry one indeed. After having made the initial discovery, after having been the first to exploit it industrially, after having developed new classes of dyes, the initiative was lost; for half a century

Germany dominated the synthetic dyestuffs industry. More serious still, Germany eventually led the world in a host of chemical industries which sprang from the making of dyestuffs, such as the manufacture of synthetic drugs. The reason for this was that the dyestuffs industry provides an excellent training ground not merely for dyestuffs chemists but for organic chemists of all descriptions.

Within a few years many of the best chemists who had studied under Hofmann in London found their way to Germany, whither he had himself returned in 1865. They included Caro, Martius, and Witt, discoverer of the important dyes known as tropaeolins. Perkin himself retired from business in 1874 and was followed by other British dyestuffs manufacturers. Outstanding among the surviving manufacturers was Ivan Levinstein who came to Manchester in 1864 and set up a factory at Blackley, making magenta, Bismarck Brown, and Blackley Blue. By 1914, when the First World War broke out, Levinsteins were the biggest dyestuffs manufacturers in Britain.

Various reasons have been put forward to account for the way in which Germany so completely outstripped Britain in the development of the industry for making dyestuffs and organic chemicals generally. Chief among these was undoubtedly Britain's own patent laws, which permitted foreigners to patent processes in Britain without making any attempt actually to work them there; it was in this that lay the great value of Caro's patent for alizarin. Another reason was the lack of duty on dyes imported into Britain, whereas British dyes imported into Germany were heavily taxed. There was also in Britain a heavy tax on industrially used alcohol; in some cases the duty was so high that it exceeded the whole of the rest of the cost of making a dye. No less important was the shortage in Britain of certain chemical intermediates essential for dye-making; of these the most important was oleum, already referred to. The teaching of organic chemistry in Germany was far superior to what it then was in Britain, so that the Germans had far better sources of technically trained men. Finally, the Germans had a far better sales organization, backed by what we now call technical service departments which were able to give detailed and authoritative advice to the purchasers and users of any of their products.

The growth of the American chemical industry, now the greatest in the world, dates also from this great expansion of the industry in Germany. Small scale manufacture of chemicals was, of course, established in America at an early date but the heavy chemical industry really began about 1876. Chemical factories, particularly for the manufacture of fertilizers, were similarly established in Canada.

CHAPTER SIX

The Chemical Industry and the First World War

As a nation Britain had been lamentably backward in encouraging science and the technology based upon it. By the end of the nineteenth century the position was, in this respect, not very different from what it had been at the beginning when it had been rightly said: 'What Minister in Great Britain ever attempted to cherish the sciences, or to reward those who cultivate them with success? . . . While in every nation in Europe science is directly promoted, and considerable sums are appropriated for its cultivation, and for the support of a certain number of individuals who have shown themselves capable of extending its boundaries, not a single farthing has been devoted to any such purpose in Great Britain.' It is true that for a short time in the nineteenth century the personal interest of the Prince Consort promised a change in this attitude, but unhappily his early death ended these hopes. While it is true that the position in Britain was rather worse than elsewhere, it was nowhere as good as it should have been – not even in Germany. No government in the world had, in 1914, anything like a clear idea of the immense practical importance of either science in general or chemistry in particular.

The outbreak of war at once revealed grave deficiencies in the constitution of the British chemical industry. Most particularly it showed, as we have just seen, an almost complete dependence on Germany for dyestuffs and many other organic chemicals and a lack of oleum for making these materials at home. The shortage of oleum had an even more serious implication, for it was of the greatest importance for the manufacture of the organic high explosives, such as picric acid and TNT, essential in modern warfare.

Another serious deficiency was in the fertilizer field; this too was a particularly regrettable branch of the industry in which to fail, for like so many others it was one founded primarily on British research. The fertilizing value of certain materials, such as animal manure, lime, and bones in improving the yields of agricultural land was well known in farming circles at a very early date, but not until the experiments of John Bennet Lawes in the middle of the nineteenth century was this purely empirical knowledge put on a logical basis. On his large estate at Rothamsted, which he inherited in 1834, he carried out extensive experiments on the mineral requirements of a large range of crops. Among the new knowledge gained was that phosphates have a remarkably stimulating effect on root crops. Following a patent taken out in 1842 Lawes set up as a manufacturer of superphosphate, made by treating natural calcium phosphate with sulphuric acid. This process converts the phosphate into a comparatively soluble form. By 1870 no less than 40,000 tons of superphosphate was being manufactured annually in Britain. Nitrates, too, imported from the immense natural deposits in Chile became of increasing importance in agriculture, so much so that in 1898 Sir William Crookes, in a famous presidential address to the British Association, warned the Western world that although widespread famine could be avoided only by still more intensive use of nitrogenous fertilizers, the Chilean deposits would be exhausted within a generation if there was no slackening of the inroads made upon them. Germany alone seriously heeded this grave warning; there the Haber-Bosch process was perfected, by which the nitrogen of the air and hydrogen derived from water were converted into ammonia, which could then be converted into nitric acid. It should, however, be mentioned that in Norway a nitrogen fixation process was developed in which air was blown through an electric arc. Unfortunately this proved uneconomic, compared with the Haber-Bosch process, even with Norway's cheap electric power.

By thus deriving her nitrogenous fertilizers from the universally abundant materials air and water Germany became completely independent of imported nitrates. It is reliably reported that had the Haber-Bosch process been fully worked out in 1913 Germany would have gone to war in that year.

Britain's position was very unsatisfactory; the more intensive cultivation of her land became urgent, but at the same time the German submarine blockade became more serious. Fortunately the menace of the submarines was overcome in time, but it is now no secret that at certain stages of the war Britain's position was critical. Not until 1917 did the British Government start to investigate experimentally the Haber-Bosch process and in the event a plant for working this did not come into operation until five years after the end of the war.

The immediate problem of the chemical industry in 1914 was to make available as quickly as possible the vast quantities of high explosive needed by the continental armies. Although two world wars have made almost everybody familiar with the general properties and effects of high explosives – and these are indeed to

some extent outmoded by atomic explosives – the
situation was very different in 1914. Up to that time
war had been waged for the most part with much less
powerful explosives of the gunpowder type, first discovered in the thirteenth century. As early as 1846
Schönbein had discovered that a powerful explosive
could be made by treating cellulose with nitric and sulphuric acids, but for many years attempts to exploit
this discovery industrially were disastrous. Sobrero
discovered that nitro-glycerine was also a powerful but
dangerous explosive. Not until the great Swedish
chemist and industrialist Alfred Nobel learnt how to
modify the violence of the explosion by absorbing
nitro-glycerine in a fine earth called kieselguhr – thus
making dynamite – could high explosives be manufactured and handled in reasonable safety. Once the
industrial hazard had been overcome the manufacture
of nitro-glycerine expanded very rapidly. New types of
nitro-explosives were also made. By 1914 Nobel's factory, built at Ardeer in 1871, had grown to a considerable size and was manufacturing dynamite, nitro-
glycerine, and cordite. The necessary oleum was made
on the spot by the contact process. There were other
smaller manufacturers, both Government and private,
but nobody had anything like the facilities which were
needed in 1914. Picric acid, made by nitrating carbolic
acid, was the chief explosive used for filling shells; TNT
(trinitrotoluene) had just been officially accepted, but
its manufacture was only just being established and
amounted to no more than 20 tons weekly. In the course
of the war no less than 238,000 tons of TNT were required. Government factories for TNT manufacture
were established as quickly as possible; the first came
into operation at Oldbury in 1915 and had a capacity of

250 tons weekly. Another great works was built at Queen's Ferry. The supply of toluene for meeting this vast output presented great difficulties, however. A certain amount was extracted from coal-tar, but even the most intensive exploitation of this source failed to meet more than a small part of the need. Before the war Dutch petroleum refineries were producing large quantities of toluene for the German dye industry; the Dutch Government, however, anxious to preserve its neutrality, refused to export toluene to Britain for the manufacture of military explosives. To overcome this difficulty, the drastic step was taken of transporting the entire refinery, owned by the Asiatic Petroleum Co., from Holland to a site near Bristol.

While the vitally important TNT manufacture was being established the manufacture of picric acid was multiplied many times over, rising to 32,000 tons yearly in 1917. As lack of toluene held up TNT manufacture, so lack of carbolic acid was one of the limiting factors in the making of picric acid. Much was derived from coal-tar. Brunner, Mond and Co, whom we have already met as manufacturers of soda by the Solvay process, directed their activities into quite different channels and produced considerable quantities of synthetic carbolic acid; the French, too, paid particularly close attention to synthetic processes.

The production of oleum in Britain by the contact process was, in 1914, on a very limited scale. While new plants were erected – primarily on systems developed in Germany – the gap was to some extent filled by importation from America. Shipping difficulties proved very serious, however. The steel drums in which the oleum was contained were sometimes not strong enough for the purpose and burst open; others were made from a grade of steel not sufficiently resistant to the highly corrosive acid. Whatever the cause of failure in the drums the result was the same; the acid got loose in the ships' holds and at once began to dissolve the plates of the hull. The construction of the new contact process plants too – at Bristol, Queen's Ferry, Gretna, and Greenwich – was beset with similar difficulties. For their satisfactory construction blocks of a special type of lava were required, but the French Government, in equal straits, had already acquired all available supplies. In consequence only second-rate and unsatisfactory building materials were for some time available.

Under the dire needs of the war ammonium nitrate – hitherto regarded by the War Office as an unsatisfactory explosive because it tended to become damp – was pressed into service to extend the limited amounts of TNT available. This was made by a modification of the Solvay process used for making soda, the sodium chloride being replaced by sodium nitrate. In addition a factory producing 2,000 tons of ammonium nitrate weekly was built at Swindon, operating a British process in which sodium nitrate reacted with ammonium sulphate.

France, too, was experiencing similar difficulties and the putting of her chemical industry on a wartime footing, with the enemy occupying a substantial part of her soil, presented great difficulties. In these circumstances both Britain and France found the assistance of America – whose chemical industry had already grown very large – of the utmost value. From America were obtained both TNT and picric acid, as well as a wide range of other chemicals including oleum, sulphur, pyrites, and cotton – the latter being required for the making of nitrocellulose propellant explosives.

The rapid expansion of the weak dyestuffs industry would, even under normal circumstances, have proved difficult; in the stress of war, when it had to compete for many raw materials and for equipment with the explosives and other industries, the difficulties were greatly magnified. The task was great and the need pressing: no less than 80 per cent of Britain's dyestuffs had been imported as such from Germany and of the remaining 20 per cent a considerable part depended on imported raw materials. Existing companies expanded their manufacturing capacity as greatly and as rapidly as possible, but drastic Government action was nevertheless necessary. In 1915 one of the leading firms – Read Holliday and Sons – was taken over and developed, under the name of British Dyes Ltd, with the aid of public money. Later the firm of Levinsteins, which had grown greatly and acquired several smaller businesses, was amalgamated with British Dyes Ltd, the resulting combine being known as the British Dyestuffs Corporation Ltd; this manufactured not only dyes, but also explosives, at great factories in Manchester, Huddersfield, and elsewhere.

While the making of explosives necessarily took precedence over all other chemical manufacturing processes, and was closely followed in importance by the manufacture of dyestuffs and related chemicals, the chemical industry as a whole had to be expanded considerably and at the same time had to make many adaptations in its processes and methods. In general the intense competition for both labour and raw materials was beneficial, for it made for greater efficiency in working and greater economy in materials. For example, nitre cake, a by-product of the manufacture of nitric acid from nitre, was at first dumped into the sea. Very soon, however, it was disposed of to various industries in which it could be used as an alternative to sulphuric acid.

No less difficult than the material shortages engendered by the war was the shortage of chemists to develop and operate the new works. There was a serious shortage, too, of skilled labour for the industry. Not only did the total number of chemists available fall short of the need, but a considerable number had enlisted in the services. Many of the latter were recalled and arrangements were improvised for training others as quickly as possible, though of necessity it was two or three years before the programme began to show results.

The natural British talent for improvisation ultimately enabled the chemical industry, despite its initial handicap, to meet the immense burden so suddenly placed upon it. The end of the war found the British chemical industry very different from what it had been in 1914. Not only had it expanded enormously and greatly increased its material assets but, no less important, there had grown up a different mental outlook both within and without the industry. Within the industry there was a consciousness of its immense importance to the nation and pride in a difficult and unexpected task well done; equally, however, it was realized that the lack of preparedness in 1914 might well have proved disastrous. In consequence there was a general determination that such circumstances should never again be allowed to arise.

The conclusion of hostilities, although relieving many stresses, created fresh problems. Two major tasks had to be fulfilled. On the one hand, a lopsided industry, geared primarily to the production of munitions and materials needed for the conduct of war, had to be adjusted to meet the altogether different demands of a world once again at peace. On the other, the grim lesson of the submarine blockade had been learnt and there was general agreement that the industry must be reorganized in such a way as to be as independent as possible of imports from abroad, though it was realized that complete independence was impossible. These needs were more easily perceived than fulfilled, for Britain's problem was the problem of all the principal belligerents. To meet the needs of the war every country had expanded its heavy chemical manufacturing capacity beyond what could be sustained in time of peace, and, in the cutting down which was necessary, each tried, so far as possible, to maintain capacity sufficient for its domestic needs. In consequence the export markets were very different from what they had been in 1914; Germany was temporarily subdued, the principal belligerents were for the most part meeting their own requirements, and competition for exports to countries which had been non-belligerent was intense.

The immense strategic importance of the Haber-Bosch nitrogen-fixation process led to the British Government's decision in 1917 to erect such a plant; when the end came in 1918, however, no active steps had been taken to put this into effect. In the changed circumstances the government of the day decided not to go ahead with the project itself but to invite private industry to undertake the task. Accordingly, Brunner, Mond and Co. were approached and this firm elected to carry out the enterprise through a subsidiary, Synthetic Ammonia and Nitrates Ltd which took over from the Government a large site at Billingham.

The task was no easy one, for it had heavily taxed the

ingenuity and resources of the highly organized German chemical industry in the years before the war and at the conclusion of hostilities very little information was obtainable from Germany except for a brief inspection of the plant. For inspiration, however, there was the indisputable fact that the process could be worked satisfactorily, always a most powerful stimulus in attacking a new and difficult problem.

To understand the full extent of the difficulties the principles of the Haber-Bosch process must be explained. On paper it is simple enough. Nitrogen and hydrogen gas will combine to form ammonia; ammonia will combine with oxygen to form oxides of nitrogen which, in the presence of water, may be converted into nitric acid. The effective combination of nitrogen and hydrogen is not, however, readily achieved. On the one hand the reaction is never a complete one; according to the conditions more or less nitrogen and hydrogen remains uncombined. Theory, confirmed by experiment, indicates that a greater yield of ammonia will be obtained by increasing the pressure; in fact, economic working is obtained only by using pressures several hundred times greater than those of the atmosphere. The production and maintenance of such very high pressures in such great quantities of gas presents very serious practical difficulties. Raising the temperature decreases the yield of ammonia, but on the other hand at ordinary temperatures the reaction proceeds so slowly that combination between the two gases is almost negligible. To compromise between the two opposing tendencies a catalyst is used to accelerate the reaction – as in the contact process for making sulphuric acid – and the gases are kept at a rather high temperature, about 400–500° C. This high temperature further complicates the problem of handling the high-pressure gases.

When the Billingham plant was being planned pressures of the order required were quite unknown in the chemical industry outside Germany, and no chemical plant manufacturer was able to supply what was needed; accordingly, the chemical engineers working on the project had to design their own plant. It is a tribute to their versatility that not only did the plant work satisfactorily but it proved in the event to be more easily managed than many plants working at normal pressures. An immense amount of laboratory research preceded the erection of the first pilot plant. It was built at the Castner-Kellner works in Runcorn, where substantial quantities of hydrogen were available as a by-product of the electrolytic manufacture of caustic soda. By 1922 two tons of synthetic ammonia were being produced daily.

For the final Billingham plant, however, a totally different source of hydrogen had to be developed, depending upon the splitting of steam into its constituent elements – hydrogen and oxygen – by passing it through a bed of burning coke. The passage of the steam cools the coke, but it can be made to glow again by passing a blast of air through it; by appropriately alternating the passage of steam and air a mixture of gases can be obtained containing nitrogen (from the air) and hydrogen in the proper proportions for ammonia synthesis. The mixture is, however, diluted with quantities of carbon monoxide and carbon dioxide. The first of these is made to combine with additional oxygen to form carbon dioxide; in this reaction more hydrogen is liberated. Finally the carbon dioxide is removed by dissolving it at high pressure in water;

when the pressure is released the carbon dioxide bubbles out of solution – just as it does in a soda-water siphon when the handle is pressed down – and is led away to be stored in a gas-holder. The mixture of nitrogen and hydrogen is heated and compressed about 300 times and the two gases then combine together, in the presence of specially prepared iron oxide as catalyst, in tall steel towers. In the reaction much heat is liberated, so that once the plant comes into operation the high temperature necessary to maintain a satisfactory yield is automatically maintained.

A great part of the ammonia produced at Billingham is converted into ammonium sulphate, for use as a fertilizer. First the ammonia is made to combine with carbon dioxide to form ammonium carbonate; the carbon dioxide needed is produced in the course of preparing the nitrogen-hydrogen mixture. The ammonium carbonate is then made to combine with a mineral known as anhydrite, a form of calcium sulphate akin to plaster of paris; this anhydrite is obtained from immense deposits existing beneath the Billingham works. This further reaction yields ammonium sulphate and, at the same time, equivalent quantities of chalk. Most of the latter is converted into cement by heating it with clay and sand. The remainder of the chalk is converted into another fertilizer called 'nitro-chalk'; this is a mixture of chalk and ammonium nitrate.

Unhappily for Britain she was not alone in recognizing the post-war importance of this kind of nitrogen fixation process, both for strategic reasons and because of the inevitable exhaustion of the Chilean nitrate deposits. Other countries, too, erected large plants of the Haber-Bosch type; in France, for example, the Claude process was developed, operating at an even higher pressure. The result was that a few years after the end of the war there was a considerable world surplus of fixed nitrogen and great economic difficulties arose. These were overcome by a voluntary international agreement between the producers, by which each was allotted an annual export quota. In this way each country was able to keep an appropriate proportion of its plants in economic operation and give regular employment to those engaged in operating them.

Steps were taken to consolidate the great advances made in the vitally important dyestuffs industry, the shortcomings of which had been so painfully revealed. For example, the British Alizarine Co., founded in 1883 to prevent alizarine manufacture becoming an exclusively German monopoly, was asked by the British Government to extend its activities sufficiently to meet all the needs of Britain and the British Empire. In addition it was asked to establish manufacture of a wide range of other dyestuffs. For these purposes a new factory was built at Manchester. In 1919 and 1921 protective legislation was introduced to regulate the importation of dyestuffs and fine chemicals of types already made in Britain.

A notable new dye – Caledon Jade Green – was discovered in 1920 in the laboratories of Scottish Dyes Limited, subsequently acquired by the British Dyestuffs Corporation. This almost unfadeable green dye was based on a chemical known as phthalic anhydride, the manufacture of which was before the war a monopoly of the Badische Anilin und Soda Fabrik, which had accidentally discovered a cheap way of making it. In 1918, however, the monopoly was broken by the development of an alternative cheap method of manufacture in the United States.

There was a growing tendency within the chemical industry for small units to combine together, for more efficient and economical working, into larger ones. The prolonged battle between the Leblanc and the Solvay processes was a major cause of this coalescence. Eventually, as we have seen, almost all the firms engaged in soda manufacture were drawn into one of two camps; on the one hand Brunner, Mond and Co. and on the other the United Alkali Company. During the stress of the war of 1914–18 a rather similar amalgamation occurred in the dyestuffs industry – Read Holliday, Levinsteins, and a number of other firms emerged from the war as the British Dyestuffs Corporation Limited. In 1926 a further great amalgamation occurred when Imperial Chemical Industries Ltd was formed as the result of a merger between Nobel Industries, the British Dyestuffs Corporation, The United Alkali Company, and Brunner, Mond and Co. There was thus created the largest industrial organization within the British Empire. In Germany the great I.G. Farbenindustrie was formed in 1925; this manufactured many classes of chemicals as well as dyestuffs. In America the du Pont Company was formed before Imperial Chemical Industries and was a powerful competitor in world markets. France and Italy, too, formed large chemical combines.

Government interest in scientific research was much stimulated by the war. Some establishments already existed to give general scientific assistance to industry; in Britain, for example, the National Physical Laboratory (N.P.L.) modelled on Germany's Physikalisch-technische Reichsanstalt of 1883, was opened at Teddington in 1900. Its primary purpose was to test and standardize materials and instruments; to-day, in very greatly extended buildings, its researches extend to metallurgy, aerodynamics, electricity, and other branches of physics and engineering. By 1915, however, the need for more extensive government aid to industrial scientific research had been recognized by the foundation of the Department of Scientific and Industrial Research (D.S.I.R.), which now maintains a considerable number of separate research institutions. The one of the greatest interest to the chemical industry is the Chemical Research Laboratory, established at Teddington in 1925; many other D.S.I.R. laboratories, for example, the Pest Infestation Laboratory at Slough, are also carrying out a good deal of chemical research.

The research done at the Chemical Research Laboratory, as at all D.S.I.R. laboratories, is of a very general nature. That is to say, it concerns itself much less with the problems of specific manufacturers than with problems of general industrial interest. It makes general studies, for example, of methods of chemical analysis, of preventing the corrosion of metals, and of economizing in the use of fuel. The more detailed needs of the chemical and other industries are met by what are called the Research Associations, maintained jointly by the D.S.I.R. and private industry. These cover such widely differing industries as those manufacturing springs, bread and confectionery, cotton goods, leather, and so on. The chemical industry as such has no Research Association directly linked with it, but the diversity of the uses to which its products are put means in fact that it is more or less interested in the Research Associations of almost all other industries. Leather manufacture, for example, involves the use of a variety of chemicals – lime for preparing hides, dyes for giving colours more attractive than that of natural leather, and salts, such as those of chromium, for making specially tough leather for the soles of boots and shoes and for similar purposes.

Similar government interest in science was expressed in many other countries. In France, where Pasteur had complained bitterly of the lack of interest in science which had followed the relatively well-endowed first decades of the nineteenth century, the end of the war saw the foundation of a counterpart of the D.S.I.R. – the Centre National de la Recherche Scientifique (C.N.R.S.). In smaller countries the effort was appropriately directed into special fields – such as wood chemistry in Sweden and hydromechanics in Holland – rather than into scientific research generally. In the United States the wartime National Research Council was put on a permanent footing; this was, however, essentially a co-ordinating organization to link up the various industrial and university research laboratories. Only in Germany was a reversal of this tendency to be seen. In 1911, in support of the belief that 'armed defence and science are the two strong pillars to support the greatness of Germany' the Kaiser-Wilhelm-Gesellschaft and its associated research institutes (now the Max Planck Institutes) were founded. These paid rich dividends, but under the Nazis came neglect and a catastrophic collapse, with the result that eventually Germany was technically completely outclassed by Britain and the United States.

This stimulation of government interest in scientific research of importance to industry is, however, only a part of the story. Industry, for its part, enormously increased its expenditure on research. In the nineteenth century most chemical works had small laboratories for carrying out tests necessary for the day-to-day working of the various plants but scarcely anything was done in the way of pure research. Knowledge of new discoveries made in university and other laboratories outside the industry found practical application only slowly. Not until 1891 were the first research laboratories established by the British chemical industry. The very first was probably the Widnes laboratory of the United Alkali Company, conceived by Dr Hurter. This example, quite startling at the time, was slowly followed by other firms. The war of 1914–18, demanding much research to establish new processes and to improve the efficiency of old ones, greatly accelerated the change, and the British chemical industry as a whole emerged with faith firmly established in the value of research of a fundamental kind. The larger firms established their own extensive research organizations; the smaller ones benefited by work done in D.S.I.R. laboratories or co-operated to establish Research Associations appropriate to their particular needs. Some of the larger firms also devoted large sums of money to enable fundamental research to be carried out in university laboratories, rightly believing that anything which promoted the advancement of science as a whole must ultimately prove beneficial to the industries which depend upon the applications of science. For the universities this financial support from the scientific industries was most opportune, for it came at a time when increasing taxation was seriously diminishing the large gifts which they had been accustomed to receive from time to time from wealthy individuals.

For the chemical industry this great sharpening of interest in research was one of the most important consequences of the First World War; many very important discoveries have resulted from it. It is, therefore, useful to recapitulate how this research effort has been divided in Britain. First, and financially the most extensive, is research, both pure and applied, carried out in the industry's own laboratories by its own scientific staff. Secondly, there is research of a wide general nature – primarily on projects of national importance which nevertheless cannot economically be undertaken by any one industrial firm – carried out in government laboratories by scientific civil servants. Finally, there is research, sponsored jointly by government and private industry, to assist groups of firms too small to establish their own laboratories.

PART II: PRESENT

CHAPTER SEVEN

The Modern Chemical Industry in Britain

THE war of 1914–18 completely revolutionized the chemical industry and marked the end of an epoch. From it sprang the complete reorganization which resulted in the structure which exists to-day. The industry is now an immense organization touching on almost every phase of both industrial and private life, and it is important to try to see where its limits lie. Care in definition is necessary, for a slight variation in our terms of reference may mean including or excluding processes which in the aggregate involve many thousands of employees and represent many millions of pounds of capital investment. It must be admitted at once, however, that no strict definition of the chemical industry has ever been formulated nor indeed – in the very nature of things – can it be; it is like trying to define where, on the fringe of a great city, the town ends and the country begins.

In attempting to make a useful definition, however, we are fortunate in being able to reap where others have sown. In 1948 the President of the Board of Trade asked the Chairman of the Association of British Chemical Manufacturers for 'a comprehensive survey of the whole chemical industry, to show its long-term plans and probable lines of development'. No such survey could, of course, be conducted without first deciding, however arbitrarily, what constitutes the chemical industry. Many official and semi-official documents were referred to but these were to a considerable extent contradictory. Among them were:

The 5th Census of Production, 1935;

the Trade and Navigation Accounts;

the Wartime Essential Work Orders;

the Partial Census of Production, 1946;

the Association of British Chemical Manufacturers' own definitions of its various groups;

the Standard Industrial Classification of the Government Inter-department Committee of Statisticians; and

the list of chemical manufactures put forward at the I.L.O. conference in 1948.

After much discussion of these perplexing documents and other relevant evidence agreement was reached. It was agreed that the chemical industry should be regarded as including the making of 'heavy chemicals, industrial gases, fertilizers, dyestuffs, medicinal and other fine chemicals, explosives, plastics, and synthetic resins, but not the compounding of chemicals to make such products as paints, insecticides, sheep and cattle dips and pharmaceutical preparations'. This last limitation is best made clear by means of an example. A proprietary medicine may contain several ingredients – for example aspirin, phenacetin, and codeine. The making of each one of these ingredients is considered a chemical process, but the mixing of them together to form a tablet is not.

For its own purposes the Association of British Chemical Manufacturers has classified the various firms falling within the agreed definition into the groups shown in the accompanying table (p. 90). The figures in this table convey concisely a great deal of information about the relative importance of the various products and the way in which the industry is now developing. In the first place it is seen that the formidable turnover of £(m)347 is being increased to no less than £(m)570; in terms of quantity this represents an increase in output from 9.8 million tons to 14.2 million tons. Although a general expansion of the industry is occurring the groups in which this increase will chiefly occur are very significant; they are in the main the groups consisting entirely or mainly of organic chemicals. Heavy organic chemicals, for example, are being increased by 117 per cent; plastics by 91 per cent; dyestuffs by 37 per cent. The greatest single increase, however, will be in the manufacture of chemical fertilizers other than those based on phosphorus or nitrogen; in effect this means largely fertilizers containing potash. This group will expand by almost 300 per cent, a significant reminder of the supreme importance of the production of food. This increase is appropriately to be accompanied by a 38 per cent increase in the production of pest-control chemicals such as insecticides and weedkillers.

It may be remarked that this great expansion will be of great value in improving the balance of Britain's foreign trade. Not only is it hoped to increase the export trade in chemicals to the extent of £(m)47 annually, but imports may be cut by £(m)28 annually; a potential overall gain of no less than £(m)75.

The report from which the above figures are culled also contains interesting statistics of the number of people employed in the chemical industry and their distribution between the various parts of it. In 1948 the

Group	Annual tonnage (1948)	Annual tonnage on completion of all development schemes	Turnover (in £(m) (1948))	Turnover on completion of all development schemes (in £(m))
1. Sulphuric acid	1,476,000	2,134,000	9.5	14.0
2. Alkalis (including chlorine)	1,759,000	2,611,000	22.1	32.1
3. Other inorganic acids and salts (including carbide)	2,561,000	3,354,000	52.1	75.8
4. Industrial gases	—	—	8.6	11.4
5. Nitrogen fertilizers	920,800	1,183,000	10.8	13.7
6. Soluble phosphate fertilizers	1,299,000	1,453,000	7.8	9.3
7. Other chemical fertilizers	172,000	710,000	2.6	10.2
8. Heavy organic chemicals	848,000	1,688,000	62.6	136.6
9. Chemicals for pharmaceutical and veterinary products	—	—	32.1	56.1
10. Miscellaneous fine chemicals	—	—	14.8	23.0
11. Dyestuffs and intermediates	195,700	237,900	38.3	52.6
12. A. Coloured pigments	41,271	53,152	6.7	8.5
B. Other pigments (including white and red and orange lead)	268,690	333,990	21.7	27.1
13. Explosives	—	—	10.4	12.7
14. Chemicals for pest control and for agricultural and horticultural purposes not elsewhere included	86,415	96,812	6.8	9.4
15. Plastic materials and synthetic resins	185,738	339,189	37.7	71.9
16. Miscellaneous chemicals	32,826	56,978	2.7	4.9

chemical industry, as previously defined, employed a total of 142,000; of these 102,000 were operatives, 34,000 administrative employees, and 6,000 scientifically qualified people engaged in research, process development and control, and so on. The expansion scheme now in course of development will call for another 25,000 employees, so that the British chemical industry may be said to find employment for about 160,000 people, about one-sixth of them being women.

The figure of 160,000 is, however, an extremely conservative one. The Ministry of Labour, for example, takes a very much broader view of the industry than does the Association of British Chemical Manufacturers and the Board of Trade; it does not, however, seek a specific definition but wisely speaks rather vaguely of the 'Chemical and Allied Trades'. According to the latter definition no less than 429,600 workers were employed in January 1949, roughly three times the number given above. These 429,600 workers were distributed as follows:

(1) Coke ovens, chemicals and dyes, explosives and fireworks	249,400
(2) Pharmaceutical and toilet preparations, perfumery and soap, etc. ..	79,600
(3) Paint and varnish	37,100
(4) Mineral oil refining, other oils, greases, glue, etc.	63,500
	429,600

This great discrepancy results primarily from the inclusion by the Ministry of Labour of certain processes not included in the narrower definition, such as petroleum refining and the mixing and blending of finished chemicals to form such products as paints, proprietary medicines, and cosmetics. Nevertheless, this Ministry of Labour definition is by no means unreasonably wide and we may therefore state with authority that the chemical and allied trades give employment to no less than half a million of Britain's population.

The Report on the Chemical Industry (1949) also contains interesting information on the sizes of the various firms concerned, which total 268. Of these only six employ more than 2,000 workers, but these six between them absorb nearly two-thirds of all the workers in the industry. One-eighth of all workers find employment in firms employing 1,000–2,000. At the other end of the scale there are 160 firms with less than 100 employees each; these account altogether, however, for only 4 per cent of all the workers in the industry.

Estimates of the total capital of the British chemical industry are exceedingly difficult to make, especially in to-day's changeable conditions. Many firms, for example, not only manufacture chemicals but carry on other quite different operations and, from the accountancy point of view, it is difficult to distinguish how much capital is associated with these different kinds of operations. Furthermore some firms have assessed the value of their plants at 'book values', i.e. the original cost of purchase. As much plant is pre-war these estimates are often quite unrealistic. In 1950, for example, Imperial Chemical Industries Limited carried out a complete reassessment of the value of their physical assets and set this at £(m)179; previously, however, basing their estimate primarily on book values, physical assets had been valued at only £(m)68. An estimate that the capital employed by the British chemical industry in 1948 was £(m)231 is now generally agreed to be much too low. It is now believed that the correct figure is in the order of £(m)150 or even £(m)250 more than this. This, as a mattter of interest, represents an invested capital of approximately £3,000 per employee.

These figures no doubt make dull reading, but they very forcibly drive home the contention that the chemical industry is not one detached from daily life but one which plays an important part in it. An industry which at the most conservative estimate employs 160,000 people, has an invested capital of about £(m)400, makes some 14 million tons of products a year, and has an annual turnover of around £(m)570 is clearly a national enterprise of the first order.

Although the A.B.C.M. classification of the chemical industry into the sixteen groups listed in the table at the beginning of this chapter is essentially arbitrary it is nevertheless convenient and logical. In the chapters which follow we shall obtain a picture of the modern industry as a whole by discussing its activities on this basis.

CHAPTER EIGHT

Sulphuric Acid, Alkali, and some other Inorganic Substances

THE historical importance of sulphuric acid and alkali was stressed in the first section of this book; their role in the modern chemical industry is no less important. Britain alone now requires for all her industries about 2 million tons of sulphuric acid, and more will inevitably be required as her industry expands. French production is about $1\frac{1}{2}$ million tons annually; in America it is of the order of 10 million tons a year.

The accompanying table indicates why sulphuric acid is such a vital factor in national economy; it is required by almost all trades, so that shortage is an immediate brake on industrial productivity generally. It is for this reason that, both in Britain and elsewhere, the sulphuric acid industry is undergoing a considerable reorganization, particularly in its choice of raw materials.

TABLE SHOWING CONSUMPTION OF SULPHURIC ACID IN GREAT BRITAIN IN THE SECOND QUARTER OF 1953

Trade Uses	Tons 100% H_2SO_4
Accumulators	2,347
Agricultural purposes	450
Bichromate and chromic acid	3,052
Bromine	4,120
Clays (fuller's earth, etc.)	2,226
Copper pickling	911
Dealers	2,755
Drugs and fine chemicals	3,234
Dyestuffs and intermediates	15,752
Explosives	7,761
Export	542
Glue, gelatine, and size	134

	Tons 100% H_2SO_4
Trade Uses	
Hydrochloric acid	13,035
Hydrofluoric acid	2,708
Iron pickling (including tin plate)	22,818
Leather	882
Lithopone	3,068
Metal extraction	996
Oil refining and petroleum products	16,054
Oils (vegetable)	2,558
Paper, etc.	1,403
Phosphates (industrial)	399
Plastics, not otherwise classified	5,336
Rayon and transparent paper	55,322
Sewage	2,766
Soap and glycerine	7,088
Sugar refining	171
Sulphate of ammonia	70,617
Sulphates of copper, nickel, etc.	5,935
Sulphate of magnesium	1,344
Superphosphates	119,740
Tar and benzole	4,705
Textile uses	5,540
Titanium oxide	40,994
Unclassified	34,532
Total:	461,295

Elementary sulphur, now mined almost entirely in the United States and Mexico, is no longer available so abundantly that no thought need be given to alternative sources. The recent shortage of sulphur, during which Britain was for a time restricted to some two-thirds of her normal requirements, finally proved less serious than was anticipated, but it was a salutary warning of the danger of becoming dependent on a single source for an essential raw material. Fuller use of other sources of sulphur is therefore important in many countries. Sicilian sulphur deposits, in normal times unable to compete economically with those of the United States, are to be developed with the aid of foreign capital. More efficient use will be made of sulphur recovered from gas works, though even under the most favourable circumstances the quantity available from this source is relatively small. It is a tantalizing thought that, as sulphur-recovery schemes are only practicable where coal is consumed in large quantities, almost enough sulphur goes up the chimneys of Britain's homes and factories to meet the whole needs of industry. Small quantities of sulphur are obtainable from petroleum refineries; for example, the refinery at Fawley produces some 12,000 tons annually. Considerably more use will be made in Britain of pyrites, though the conversion of sulphur-burners to pyrites-burners is not easy.

On the Continent the sulphuric acid situation differs from that in Britain and the United States, for there has not been the same dependence upon elementary sulphur. Although sulphur has several advantages a plentiful supply of pyrites in Europe, especially in Spain, encouraged Continental manufacturers to use this almost exclusively as their raw material.

The question of utilizing anhydrite – an ore chemically related to plaster of paris – deserves more attention than we have so far given to it, for Britain and many other countries have extensive anhydrite deposits. The manufacture of sulphuric acid from anhydrite was first carried on at Leverkusen, Germany, in 1926. Since 1929 sulphuric acid has been made at Billingham from locally mined anhydrite, but the price control for sulphuric acid introduced by the Board of Trade in 1940 made fuller use of this material uneconomic. After the war, however, as a consequence of the sharp rise in the price of both pure sulphur and pyrites, the controlled price of acid advanced sufficiently to make it possible to make fuller use of anhydrite. Large new plants for the utilization of anhydrite are being built in Britain and elsewhere. In the years just after the war world manufacture of sulphuric acid from anyhdrite was about 210,000 tons annually, but when present plans are completed it will rise to about 400,000 tons.

On the other side of the ledger, efforts are being made to economize in the use of the sulphuric acid available. As shown in the table the biggest single consumers of sulphuric acid (120,000 tons quarterly) are the manufacturers of superphosphate fertilizer. Research in Britain has recently shown that nitric acid – much more readily available – can replace sulphuric acid to a considerable extent in converting rock phosphate into phosphate fertilizer. If this process can be exploited on a large scale, as seems possible, it alone can effect a considerable saving in sulphur.

The contact process of sulphuric acid production has by no means displaced the traditional lead-chamber process. Although about 70 per cent of American acid is made by the contact method, in Britain the proportion is roughly even, but slightly in favour of contact.

The manufacture of alkali has always been one of the principal foundations of the chemical industry and it remains so to-day. As a result of the various mergers which resulted from the competition between the Leblanc and the Solvay processes and the reorganization after the First World War, soda manufacture in Britain is now in the hands of a few large firms. The principal raw material of the alkali industry is salt, of which enormous quantities are consumed by the chemical industry. World production of salt is now about 25 million tons a year, and barely one-tenth of this finds its most familiar use, the flavouring and preservation of food. The remainder is transformed into such varied chemicals as soda, caustic soda, hydrochloric acid, chlorine, metallic sodium, cyanides, and glass. Most salt is mined from underground deposits, as in Cheshire, though in hot countries solar evaporation of sea water is still much used.

For certain purposes – for example, petroleum refining and soap-making – ordinary washing soda must be converted into the caustic variety. This, as we have seen, may be done by treatment with lime. In addition, however, a considerable amount of caustic soda is made by the electrolytic decomposition of salt, using Castner-Kellner cells. This process yields chlorine as a by-product; for this there is a considerable and growing demand. In Britain the extent to which the Castner-Kellner process is worked is decided by the quantity of by-product chlorine which can be disposed of.

To chlorine's original use as the raw material for making bleaching powder many others have been added. Outstanding among them is the manufacture of chlorinated organic solvents, extensively used for dry-cleaning and for degreasing machinery. Other organic chemicals containing chlorine have also assumed great importance in recent years. Vinyl chloride, for example, is the starting-material for the very widely used p.v.c. plastics; ethyl chloride is needed for leaded petrol. On the inorganic side, aluminium chloride, made by reaction between chlorine and alumina, is used for 'cracking' high-boiling petroleum fractions in order to convert them into more volatile ones suitable for use as a fuel; aluminium chloride is also necessary for making certain dyes and organic chemicals.

Hydrochloric acid, once available in embarrassing excess to those who worked the Leblanc process, is now made in considerable quantities by combining the chlorine which appears at the positive pole of the

Castner-Kellner cell with the hydrogen evolved in equivalent quantities at the negative pole. Much hydrochloric acid is also made by what is essentially the first stage of the obsolete Leblanc process, namely the treatment of salt with sulphuric acid. Additional quantities are evolved as a by-product of the chlorination of organic compounds.

As indicated above, in the electrolytic decomposition of brine the production of chlorine is comparable in importance with that of caustic soda. It may be remarked also that an allied electrolytic process of decomposing brine produces the important salt sodium chlorate. This, and the related ammonium and potassium salts, are used in the metal and explosives industries; sodium chlorate is a weed-killer.

The general expansion of the alkali industry, and of industry generally, has been accompanied by an enormous increase in the production of limestone and lime. In the United States, for example, about 160 million tons of limestone and chalk are quarried annually; of this about 6 million tons is burnt to lime. Total production figures for Britain are not available, but in 1947 quarries other than those owned by iron and steel manufacturers – the biggest consumers – yielded 17 million tons of limestone.

The many inorganic substances comprising the third group of the ABCM classification are not easily defined, for they merge almost imperceptibly into what are termed miscellaneous fine chemicals. They include a variety of inorganic chemicals made in large quantities, such as ammonia, nitric acid, calcium carbide, hydrogen peroxide, sodium perborate, salt cake, sodium peroxide, sodium silicate, and a number of others which have already been mentioned.

Of these varied products ammonia is perhaps the most important, for it finds many uses. The method of manufacturing ammonia for subsequent oxidation to nitric acid has already been described. Nitric acid is used primarily in making dyes, drugs, and explosives, all of which will be discussed later in their appropriate chapters. Very large quantities of ammonia are converted into ammonium sulphate for use as an agricultural fertilizer.

The bleaching action of chloride of lime is due primarily to its oxidizing powers and a number of other powerful oxidizing agents have been brought into use for this and other purposes. Of these perhaps the most familiar is hydrogen peroxide, which is found in most homes, being there used for the double purpose of a disinfectant and a bleach. In Britain the manufacture of hydrogen peroxide was originally established at Luton in order to be near the straw-hat factories which absorbed most of the output. Hydrogen peroxide for domestic use is usually sold under the rather puzzling title 'twenty volume'; this means simply that one volume of the solution will, on complete decomposition of the peroxide, yield twenty volumes of oxygen gas. It is a particularly convenient oxidizing and bleaching agent, because the only products formed in its action are water and oxygen gas, both completely innocuous and easily disposed of.

Hydrogen peroxide is made by treating barium peroxide with phosphoric acid – two chemicals not previously mentioned. The starting point for barium peroxide is a mineral called barytes. This is found in many parts of the world, including Ayrshire, Devonshire, and Shropshire. If barytes is heated with coke it loses its oxygen and is converted into the sulphide; treatment of this with soda causes barium carbonate to be precipitated. The barium carbonate is converted on heating to barium monoxide in the same way as calcium carbonate is converted into lime (calcium oxide). On further heating in air, however, the barium monoxide takes up more oxygen to form barium peroxide.

If this peroxide is still more strongly heated it loses this extra oxygen and is reconverted into the monoxide. This addition and loss of oxygen was once widely used in Britain for the preparation of pure oxygen; the process was worked by Brin's Oxygen Company, now the British Oxygen Company Limited.

Pure phosphoric acid – the other chemical needed for making hydrogen peroxide – is now made from elementary phosphorus by treating it with super-heated water in the presence of a catalyst. Phosphates find many uses, for example in baking, in fire-proofing wood, and in softening water. Phosphates for agricultural purposes are of much lower purity and are made from various phosphate rocks.

The making of elementary phosphorus itself is an interesting and important process. The first large-scale industrial method was developed about 1844 by Arthur Albright of Birmingham, who supplied it to the match industry, to which he was already supplying potassium chlorate. The phosphorus originally made – by treatment of phosphates, such as bones, with sulphuric acid and carbon – was what is called yellow phosphorus; this caused much suffering in the match industry as the workers suffered severely from the fumes. In 1849, however, Albright acquired from Schroetter, a Viennese chemist, the rights in a process for making a new and comparatively harmless variety of phosphorus – red phosphorus. The existence of phosphorus in these two forms, red and yellow, is known to the chemist as allotropy; a more familiar example is the existence of carbon in several forms such as graphite and diamond. The rapidly expanding Swedish match industry placed large orders for red phosphorus, so large indeed, that Albright, a staunch Quaker, at first refused the orders, believing that such quantities of phosphorus could be required only for military purposes.

Phosphorus is to-day made almost entirely by an electrical method in which a heavy current is passed through a mixture of phosphate, carbon, and silica. Phosphorus compounds of importance in the modern chemical industry include, apart from those already mentioned, the chlorides and oxychlorides – much used in organic syntheses – and calcium phosphide, used in self-igniting marine flares.

From hydrogen peroxide is made another important bleaching agent – sodium perborate – which can be made by reaction between hydrogen peroxide and a mixture of borax and caustic soda. Sodium perborate is an ingredient of many household washing powders and is also used in toothpastes, gargles, and other pharmaceutical products.

The production of sulphur dioxide gas for the making of sulphuric acid has already been mentioned. This gas has, however, other important uses, for from it are derived, by reaction with alkali, the salts known as sulphites. The simplest of them is sodium sulphite, used as a mild bleaching agent for wool and silk, for neutralizing excess of bleaching powder when the latter is used, and in making photographic developers. A related, slightly acid, salt is sodium bisulphite; hot solutions of this have the power of dissolving gummy products from wood and it is very extensively used for making wood-pulp for paper-making. If treated with zinc dust the bisulphite yields the hydrosulphite, used in large quantities in the sugar and dyestuffs industries because of its 'reducing' properties, reduction being, in the chemical sense, the opposite of oxidation. Finally, mention must be made of sodium thiosulphate – the photographer's 'hypo'; it is also used in the textile industry for removing the last traces of bleach, and in the extraction of silver.

The final member of this rather complex and varied group is one of the latest arrivals in the chemical industry – the element fluorine. Chemically this is closely related to chlorine. The history of fluorine goes back almost two centuries, for its existence was recognized by Davy at the beginning of the nineteenth century; the element itself is, however, so violently reactive that all attempts to isolate it led merely to its exchanging one partner for another. The first to isolate fluorine was the French chemist Moissan who obtained the green gas in 1886, using an electrolytic cell made of the highly resistant metal platinum. Moissan explored the chemistry of fluorine in some detail, but it proved so difficult to handle that even in chemical laboratories only a few specialists worked with it. Patient experiment established, however, that the compounds of fluorine were sufficiently interesting in their properties to justify an attempt to handle this intractable element on an industrial scale and fluorine-producing electrolytic cells have been developed in recent years in Germany, Britain, and the United States. All are based on the electrical decomposition of a solution of potassium fluoride in hydrofluoric acid. Research has shown that under certain conditions metals less expensive than platinum may be used; these include copper and certain types of steel.

Hydrocarbons in which some or all of the hydrogen has been replaced by fluorine, or partly by fluorine and partly by chlorine, have exceptional advantages as circulating fluids in refrigerating systems and are widely used for this purpose. Unlike their parent element fluorine they are quite unreactive and they are not poisonous. Dyestuffs containing fluorine show important differences from their counterparts containing chlorine. Fluon, a fluorine-containing plastic, has great technical potentialities, being highly resistant to both heat and chemical attack.

The industrial chemistry of fluorine is still in its infancy but as the raw material for its manufacture is abundant, and the difficulties of preparing and handling it have been overcome, further very interesting and important developments are clearly only a matter of time. With the help of fluorine new types of pharmaceuticals, plastics, solvents, lubricants, insecticides, and other products are to be expected.

CHAPTER NINE

Industrial Gases

ALTHOUGH the chemical industry makes use of many gases, some of which we have already met, the common definition of industrial gases admits only about a dozen. These are oxygen, hydrogen, nitrogen, 'inert' gases (such as neon), carbon dioxide, acetylene, propane, and butane. The last three, being of an organic nature, form a group on their own.

Oxygen, hydrogen, and nitrogen we have already encountered collectively because they are the basic materials for making synthetic ammonia, nitric acid, and related products. Oxygen has many other uses, however, and its production is an important part of the chemical industry. The old Brin process – based on a continuous alternation between barium monoxide and dioxide – has now been replaced by one based on the fractional distillation of liquid air. The present extent of the manufacture of pure oxygen is difficult to estimate, but before the war it amounted in Britain to about 300 million cu. ft annually. About three-quarters of this is used in oxy-acetylene and similar torches for cutting or welding metal. The remainder finds many uses, notably in medicine and in scientific research.

Atmospheric nitrogen is used in huge quantities for making synthetic ammonia; this is, however, its only important use. Air consists almost entirely of oxygen and nitrogen, these two gases comprising more than 99 per cent by volume. The nature of the very small residue – and indeed its very existence – was virtually unknown until comparatively recent times. As early as 1785, however, the brilliant but eccentric British chemist Henry Cavendish suspected that some 1/120th part of the air consisted of gases other than nitrogen and oxygen. Not until 1894 was the nature of this residue made clear; in that year Rayleigh and Ramsay discovered argon. A year later Ramsay discovered helium, the existence of which in the sun was already known through spectroscopic investigation, and soon afterwards three new constituents of air – neon, krypton, and xenon – were discovered. In 1901 a sixth related gas, radon, was isolated by Dorn, though not from air; it is a product of the radioactive decay of radium. Of these gases argon is relatively much the most abundant, being present in normal air to the extent of 0.94 per cent by volume. The abundance of the others is shown in the accompanying table.

Gas	Abundance in air, by volume
Neon	1 part in 65,000
Helium	1 part in 200,000
Krypton	1 part in 1,000,000
Xenon	1 part in 11,000,000

These gases are remarkable in their chemical properties – or rather in their lack of them. Unlike all other elements, which will more or less readily combine to form a range of compounds, the rare gases of the atmosphere will form no chemical compounds whatsoever; they are completely inert. In recent years they have, in consequence, found many uses and they are manufactured in considerable quantities as a by-product of oxygen manufacture or, in the case of krypton and xenon, in specially designed plants. In certain parts of the world, notably the United States, comparatively large concentrations of helium (up to 8 per cent) are found in certain natural gases which pour out of the earth's crust. Although not so light as hydrogen, helium is extensively used in the United States to fill military observation balloons, as it possesses the immense advantage of being non-inflammable. Argon, krypton, and xenon are widely used in the manufacture of electric light bulbs, their introduction into the globe permitting a higher working temperature than in a vacuum and therefore more light per unit of current. No other gases are suitable, as at the high temperatures all combine with the glowing filaments. Neon is familiar in the electric discharge lamps used in street advertisements and as night-lights; argon and helium are used for similar purposes. Liquid helium is of great value in scientific research in attaining temperatures very close to the absolute zero (−273° C.).

The use of hydrogen for the manufacture of synthetic ammonia and nitric acid is by no means its only industrial application. It is used, for example, for hardening fats and oils, primarily for margarine manufacture; in the presence of a catalyst such as nickel many fats and oils will take up considerable quantities of hydrogen and as a result are converted into new products of higher melting-point.

An outstandingly interesting use of hydrogen is in the manufacture of synthetic petrol. The increasing use made of the internal combustion engine since the beginning of this century, and particularly its strategic importance in time of war, led the governments of countries without petroleum deposits, but possessing coal, to investigate the possibility of converting coal into petrol. As might be expected in view of her long and careful preparation for the war which eventually broke out in 1914, Germany took the lead, but during that war synthetic petrol made only a slight contribution to her total needs.

British interest dates from 1921, when Dr Bergius arrived from Germany and explained his synthetic petrol process to the Department of Scientific and Industrial Research. With memories of the submarine blockade fresh in their minds officials paid close attention to his claims and in 1923 experiments were made to examine the possibility of working the Bergius process in Britain. Four years later, as a result of the personal interest of Sir Alfred Mond, the experimental work was taken over by the then newly formed Imperial Chemical Industries Limited. Although a pilot plant was soon put into operation large-scale working proved impossible for economic reasons. In 1933, however, the British Government finally realized the immense importance of the process to the nation and guaranteed a preference to all home-produced motor spirit. At once work was commenced on a £(m)3 plant at Billingham; this was opened by Mr Ramsay MacDonald in 1935 and was designed to produce 150,000 tons of petrol annually. The process is a complicated one and, like the Haber-Bosch process, demands the handling of great quantities of gases at very high temperatures and pressures. Its strategic value was proved during World War II. The raw material of the process is coal, coal-tar, or creosote; at the time of writing the latter is the most important. Plants for the synthesis of petrol from coal have been set up in the United States, South Africa, France, and elsewhere, generally for strategic reasons as the synthetic product is competitive with natural petroleum only in special circumstances.

In the years before World War II, Germany too, for similar strategic reasons, paid much attention to the manufacture of petrol from coal by what is known as the Fischer-Tropsch process. Despite the special attention given them by the Allied air forces the synthetic petrol plants proved invaluable to Germany and continued production right up to the final collapse. Current research on synthetic petrol processes is directed mainly at producing more by-products suitable for sale as raw materials for the organic chemical industry.

The last of the inorganic industrial gases is carbon dioxide. This is produced in substantial quantities as a by-product of nitrogen fixation by the Haber-Bosch process; it is also formed in lime burning, many fermentation processes, and in other ways. It finds a great many uses. At Billingham surplus carbon dioxide is converted into the solid form, which sublimes at $-78°$ C. A small fraction of the carbon dioxide produced in the fermentation industries, such as those manufacturing industrial solvents, is similarly treated. Solid carbon dioxide is for many purposes an extremely valuable refrigerant, since it does not melt to form a liquid, as ice does, but changes directly into gas which is dissipated continuously into the surrounding air. It finds many uses, most particularly in the preservation of perishable foods. It is also used by undertakers, by engineers for shrink-fitting, and in other ways.

A great deal of carbon dioxide gas is used for the manufacture of soda-water and other aerated drinks, a use originally proposed by Joseph Priestley. Carbon dioxide is also used in fire-fighting when, as in oil and petrol fires, water is unsuitable. Being a relatively heavy gas the carbon dioxide forms a blanket over the flames and thus prevents access of air.

Of the organic gases much the most important is acetylene, made by the action of water on calcium carbide; a few cyclists still use acetylene lamps in which the gas is generated in this way as required. Although discovered as long ago as 1836 little industrial use could be made of acetylene until the introduction of the electric furnace some fifty years later, for without this the large-scale production of cheap calcium carbide was not possible. In countries such as Switzerland, Norway, and Canada, where hydro-electric power is cheap, calcium carbide is made on a very large scale, mainly for export. Before the war, for example, Britain imported about 60,000 tons of carbide annually.

Treatment of carbide with water leads to the immediate evolution of acetylene gas; a residue of 'carbide sludge', which is essentially slaked lime, is left in the generators. Unless used on the spot, the gas is compressed and stored, after purification, in steel cylinders packed with porous briquettes soaked in acetone. This unusual method of storage is essential, for unless dissolved in acetone or a similar solvent compressed acetylene is liable to explode spontaneously with great violence.

In recent years acetylene has become of the greatest importance in industrial organic chemistry. In the presence of mercuric sulphate as a catalyst, for example, acetylene will react with water to form a colourless liquid known as acetaldehyde, a starting-point for many other synthetic organic chemicals such as acetic acid and acetone. Its reaction with chlorine yields a valuable series of non-inflammable organic solvents.

A large part of all acetylene produced is used in oxy-acetylene blowlamps. A mixture of acetylene and oxygen yields an intensely hot flame used for welding and for metal-cutting.

Butane, a hydrocarbon gas, is derived from a number of sources, such as synthetic petrol plants, petroleum refineries, and oil-wells. It is readily compressed to form a liquid and in this form is familiar in country districts as a substitute for coal-gas. The related gas propane is similarly available. The main use of both gases at the present time is as a convenient gaseous fuel.

Finally we must mention one or two other industrial gases of lesser importance. Nitrous oxide – only too familiar to most of us as a dental anaesthetic – is made by decomposing ammonium nitrate. After very careful purification it is compressed and liquified and distributed in small cylinders. Ethylene gas, made by decomposing alcohol by heating, is used in large quantities for manufacturing polythene and related plastics, which are of great technical importance, and for controlling the ripening of stored fruits such as apples.

CHAPTER TEN

Agricultural Chemicals

THE importance of artificial fertilizers, indispensable for the systems of intensive cultivation by which alone the immense populations of western civilization can maintain their present standard of living, has already been stressed. They fall into three main groups – those containing nitrogen, phosphorus, and potash respectively.

The most widely used nitrogenous fertilizer is ammonium sulphate. This is not absorbed as such by growing crops, but bacteria present in all normal soils are able to convert it into soluble and readily utilized nitrates. Another important British fertilizer of this type is 'nitro-chalk', a mixture of synthetic ammonium nitrate and chalk. The greater part of the nitrogenous fertilizers used in British agriculture to-day is of synthetic origin; present production is of the order of 1 million tons annually. In addition, however, substantial quantities of ammonium sulphate – totalling about 290,000 tons annually – are obtained as a by-product of the carbonization of coal in gas-works or coke-oven works. One of the products of coal distillation is ammonia, which is either removed from the gas by bubbling it through sulphuric acid or is washed out with water, forming ammoniacal liquor which is then distilled with lime and the resulting ammonia bubbled into sulphuric acid. In both cases, the product is ammonium sulphate. It is interesting to note that twenty years ago, when less coal was being coked than now,

the output of by-product ammonium sulphate was
385,000 tons; it is therefore clear that much of the ammonia produced in this way is going to waste, the
reason presumably being the efficiency and cheapness
of the Haber-Bosch and related processes.

Cyanamide is another important nitrogenous fertilizer; as, however, it is made from calcium carbide, the
manufacture of which demands cheap electric power,
it is only in countries such as Norway, which have this
facility, that it is at all widely used. Large quantities of
cyanides, some used as agricultural fumigants, are made
by heating cyanamide with salt.

Other important fertilizers are calcium and sodium
nitrates; these have the advantage in certain cases of
having an immediate effect, as they are directly absorbed
by growing plants. Cyanamide, on the other hand, has
to undergo a series of changes in the soil before plants
can use it; as some of the intermediary products are
poisonous the soil must be dressed a week or two
before seed is sown.

Phosphorus is of particular importance in stimulating the growth of roots. It is applied to the soil almost
exclusively in the form of superphosphate, made by
treating phosphate rock with sulphuric acid. British
production of superphosphate – from phosphate rock
mostly imported from North Africa – to-day exceeds
1,100,000 tons annually and this makes a very heavy
drain on available supplies of sulphuric acid. Plans are
now being made to manufacture increased quantities of
a more concentrated fertilizer of this type, known as
triple-superphosphate; this contains about 45 per cent
of useful phosphorus (expressed in terms of phosphorus
pentoxide) instead of the 18 per cent found in ordinary
superphosphate.

Another important phosphate fertilizer is ammonium phosphate which, being soluble in water, has a rapid action. Basic slag, a by-product of the steel industry, is also used in large quantities. Less important, because only small quantities are available, are ground bones. A little rock phosphate is used as such, instead of being converted into superphosphate, but the phosphorus locked up in it is only very slowly available to plants.

The last of the three major groups of fertilizers is that comprising substances containing potassium. This is particularly valuable in the growing of fruits and vegetables. Potassium minerals all contain some proportion of magnesium and sodium salts. Traces of magnesium are essential to plants – for it is a constituent of their green pigment chlorophyll – and sodium salts promote the growth of a few special crops such as sugar beet. Potassium fertilizers are therefore valuable in supplying these further essential elements. Originally the main uses of potash were in the making of glass, preparing leather, and manufacturing soap. Now, however, the balance has entirely changed and of a greatly multiplied production almost all is used in agriculture. For this purpose a number of potassium salts are used – notably the chloride, sulphate, and nitrate. Formerly, Germany had almost a monopoly in the supply of potash, for immense deposits occur at Stassfurt. Now, however, potassium minerals are mined in France, the United States, Russia, Poland, Spain, and Israel. More important, from the British point of view, is the discovery of an extension of the Stassfurt deposits beneath parts of Yorkshire, though unfortunately at great depth.

Very large quantities of lime are used for agricultural purposes but this plays a comparatively small part in plant nutrition. Much more important is its ability to

neutralize acidity in the soil, for this sometimes inter-
feres with the utilization of other minerals by crops.
Lime is produced by burning limestone or other forms
of calcium carbonate, but calcium carbonate itself is
also used to neutralize acid soils; varieties used include
ground limestone, waste carbonate from paper-works
and beet-sugar factories, and broken sea-shells.

For many purposes a general fertilizer is required
rather than a single one of any of the types described
above. A common mixture is ammonium sulphate,
superphosphate, and a potassium salt. From such mixtures lime must, however, be excluded, as it decom-
poses ammonium salts.

Recent research has shown that for healthy growth
plants require in addition to substantial quantities of
nitrogen, phosphorus, and potassium, traces of several
other elements; the importance of sodium and magnesium in this respect has already been mentioned.
Shortage of boron, for example, is occasionally the
cause of rot in turnips and beet; it can be corrected by
treatment of the soil with borax. In reclaiming certain
types of moorland, soil treatment with a copper salt –
usually the sulphate – is helpful; in tropical agriculture
deficiency of zinc in the soil crust must sometimes be
corrected with salts of the metal.

Although this is not the place to discuss in any
detail the well-known controversy between the use of
natural fertilizers – which in fact means one form or
another of farmyard manure or of compost – it may
perhaps be remarked that this arises from a misunder-
standing of the issues involved. Natural manures serve
two almost equally important purposes – they provide
nutrients essential for the growth of plants and, if
properly used, they maintain the soil in the physical

condition necessary for growth; for example, they lighten heavy soils and so permit young roots to push through them, and they cause light soils to retain the water essential for growth. Synthetic fertilizers can fulfil only the first of these purposes; they have as a rule little or no effect on the tilth of the soil. Consequently chemical fertilizers must be used in conjunction with methods of keeping the soil in proper condition, though recently synthetic materials have been developed for altering the texture of the soil. As, however, no responsible advocate of synthetic fertilizers has contended otherwise, those who seek to condemn them are creating a purely artificial dispute. Furthermore, it is to-day physically impossible to use only natural fertilizers, for these are insufficient to meet the demands of modern methods of intensive cultivation, which synthetic fertilizers alone make possible.

The debt of modern agriculture to chemistry is, however, by no means limited to the production of fertilizers. Fertilizers promote the growth of crops but it is also necessary to destroy the enemies – primarily weeds and insect pests – which attack them during growth. In Britain alone crop losses from these causes exceed £(m)50 and world losses are immensely greater.

The earliest insecticides were of vegetable origin. Nicotine preparations, made by steeping tobacco leaves in water, were used as early as the eighteenth century. At the end of the nineteenth century derris preparations, made from the roots of certain tropical shrubs, were introduced; these are extremely rapid and deadly in their action. Mineral insecticides, based chiefly on the poisonous compounds of copper and arsenic, were introduced in the nineteenth century. At the same time the rapidly expanding coal-tar industry offered new materials such as naphthalene and creosote. Other insecticides now in general use include hydrogen cyanide and thiocyanates (sulphur-nitrogen-carbon compounds). Although quite effective, these substances present certain practical difficulties because they are toxic not only to insect pests but also to human beings and domestic animals; great care is therefore necessary to see that food is not contaminated and that operators handling the agents are not injured.

The ideal insecticide is one which destroys insects but is innocuous to other animals; in recent years this ideal has almost been achieved. DDT, a Swiss discovery exploited by the Western allies during World War II, is very deadly – although slow in its action – to insects, but not very poisonous to human beings and animals. In 1942 the British chemical industry developed a somewhat similar type of insecticide known as Gammexane; this is a chlorine derivative of benzene. Both DDT and Gammexane are applied in various forms, but primarily as dusting powders and sprays. They have proved immensely successful against a host of insect pests throughout the world – the boll weevil of cotton crops; the Colorado beetle; wire-worm; the cabbage white butterfly; and many others. Most important of all, perhaps, their use has pointed the way to control of locusts, the most destructive of all pests in tropical agriculture.

These new insecticides are of great significance in human and animal health. Many very serious diseases, such as malaria and typhus in man and ngana in cattle, are spread solely by insect vectors such as mosquitoes, lice, and tsetse flies.

The accumulation of great stocks of grain and other foods – for strategic and other reasons – offers a new and important field of application for insecticides, as they present particularly vulnerable targets. For the fumigation of grain a volatile insecticide – methyl bromide – is to-day very widely used; this was developed in the United States.

To the farmer and market gardener, weeds, which stifle the growth of crops both by cutting off sunlight and by depriving the soil of nutrients, are scarcely less serious a menace than insect pests. In Britain alone, for example, loss of crops due to weeds is estimated at more than £(m)16 annually. Although weeds can to a considerable extent be controlled by mechanical methods, such as hoeing, very large quantities of weed-killers are also used. Until recently dilute sulphuric acid was one of the most widely used weed-killers. If applied to corn crops at the correct time the growing plants, having erect narrow leaves with a waxy protective coating, are almost unharmed, but associated weeds, with broad leaves on which the acid accumulates, suffer severely. DNOC (dinitro-*o*-cresol) and sodium chlorate are also widely used as weed-killers.

Recent research, however, has resulted in the development of important weed-killers of a totally new kind. Plants, like animals, depend for their healthy growth on the production in their tissues of tiny quantities of certain rather complicated organic substances known as hormones. The amounts of these present must, however, lie between certain limits; excess causes very abnormal growth, with fatal results. Some substances possessing effects similar to those of the natural plant hormones are synthesized industrially. They have a selective action, being very much more active against the class of plants known as dicotyledons – to which most weeds belong – than to the monocotyledons, of which grass, wheat, rye, barley, and oats are representative. Very small quantities of these substances have remarkable results. As little as two pounds per acre will often suffice to keep all weeds in a corn crop under control. Great attention was paid to these substances in Britain during World War II. When the first of these, the British-discovered Methoxone, made its debut the Ministry of Agriculture was so impressed that an immediate order was placed for a very large quantity of the dilute form used for field application. These trials were spectacularly successful. More recently discovered substances have the power of attacking monocotyledons in preference to dicotyledons.

Synthetic plant hormones of this type are used for other purposes, particularly by fruit growers. If sprayed on apple-blossom in carefully controlled quantities premature drop of fruit is avoided later in the season; tomatoes and other fruits can be made to set without pollination; cuttings and seedlings can be induced to root much more freely than is normal.

Apart from insect pests and weeds, growing crops are liable to attack by a considerable range of fungi. When these have once established themselves they are difficult to eradicate, though certain types of spray are useful; prevention is, however, a good deal easier. Particular attention was paid to this problem in the wine-growing regions of France and various fungicides based on sulphur were developed there; the well-known Bordeaux mixture is a mixture of copper sulphate and milk of lime. Potato blight is similarly treated. The origins of many fungus diseases lie latent in seeds before they are sown and it is now common practice to treat seeds with a fungicide before sowing; the substances used are usually organic compounds of mercury.

CHAPTER ELEVEN

Organic Chemicals

THE range of organic chemicals is so great that for convenience it is divided into two groups – heavy and fine. Heavy chemicals are essentially those produced in bulk and used in large quantities; fine chemicals are made on a comparatively small scale, some indeed in quantities of only a pound or two. The division is, however, purely arbitrary and one group overlaps the other. The difficulty of defining fine chemicals is indeed such that, in despair, a president of the Society of Chemical Industry said that 'a fine chemical is something made by a fine chemical manufacturer'.

The most important of the heavy organic chemicals are the alcohols and their derivatives. To the layman alcohol usually means only one thing – the essential ingredient of a very large number of pleasant drinks. While in his leisure hours the chemist appreciates this meaning no less than his fellows, his professional relationship with alcohol is very different. To him it is not one substance but a group of substances; not an intoxicating principle of liquor but a vitally important raw material of the organic chemical industry.

The principal alcohol of drinks such as beer, wine, and spirits is distinguished chemically as ethyl alcohol. These intoxicating drinks are all made by fermenting sugary or starchy materials such as barley, grape-juice, honey, or potatoes with yeast. Similar fermentation processes are used for the industrial production of ethyl alcohol, though the emphasis is then, of course, entirely on producing the maximum possible concentration of alcohol in the fermented liquor. One of the most important raw materials for alcohol manufacture is molasses, the thick brown syrup, containing up to 50 per cent of sugar, formed as a by-product of the production of cane- or beet-sugar. It is diluted with water, warmed slightly and seeded with a selected strain of yeast. The main fermentation takes place in huge tanks of perhaps 75,000 gallons capacity. These require very careful design and preparation, as it is necessary to see that the fermenting liquor does not become infected with the stray yeasts and bacteria which always float about in the atmosphere. These organisms also can break down sugar and starch, but often with the formation of products very different from the ethyl alcohol which is desired. When fermentation is complete, usually after about two days, the liquor, which by then contains up to 8 per cent of alcohol, is distilled in a special type of still; the ultimate product contains 96 per cent of alcohol and 4 per cent of water. For technical reasons the removal of the last traces of water is extremely difficult, but for almost all purposes this is not necessary.

The fermentation process is accompanied by a brisk evolution of carbon dioxide gas, making the liquor bubble and seethe. Some of this gas, formerly allowed to escape into the air, is now collected and compressed into solid form for use as a refrigerant.

Huge quantities of alcohol are now used in industry as a solvent or as a stepping-stone to other chemicals, but its use for this purpose is complicated by its intoxicating properties. In its pure form it is a dangerously poisonous substance which could not in any event be made freely available to the public. Moreover in most countries alcohol is an exceedingly heavily taxed commodity and, once this source of revenue has been established, even the pressing needs of chemical industry would not persuade national treasurers to allow tax-free alcohol to find its way to the public. To compromise between these two opposing claims – of the public to alcoholic drinks on which it is prepared to pay an enormous duty and of the chemical industry to an important raw material which must be cheap – alcohol for industrial use is diluted slightly with highly unpalatable substances such as wood naphtha or pyridine. Sometimes some violet dye is added as well. Alcohol so treated – known as methylated spirit – is not normally liable to tax and for most purposes the presence of the added denaturants is not harmful. For certain medical and scientific purposes, and a few industrial ones, untreated alcohol is available in Britain under strict supervision at a reduced rate of duty. The struggle of the British chemical industry for cheap alcohol was, however, a very long one, and in the long run the near-sitedness of the Treasury cost the country dear; not until as recently as 1946 did it come on an equal footing with American and Continental countries.

The uses of ethyl alcohol are so numerous that only a very few can even be mentioned here. It is used in motor fuels; in some countries the addition of a considerable percentage of alcohol to all petrol is obligatory by law. It is used also as a solvent for making many polishes and varnishes, for the preservation of biological and medical specimens, in embrocations and other pharmaceutical and toilet preparations, as a fuel in small stoves, in ships' compasses, for making printing inks, and for many other purposes.

Ethyl alcohol is also the source of an immense range

of other chemicals, including acetone, acetaldehyde, acetic acid, and ethylene gas. As each of these many chemicals may itself be the source of a series of others it will be apparent why ethyl alcohol is one of the key substances of the organic chemical industry. The details of these complex chemical transformations need not concern us here, but it may be remarked that ethyl alcohol is an essential ingredient in making such diverse products as anaesthetics, dyes, artificial silk, certain plastics, and many fine chemicals.

The measure of the industrial importance of alcohol is given by the fact that in Britain alone annual consumption for all purposes is of the order of 50 million gallons annually. At the height of the Second World War the United States was using – for a process now no longer employed – no less than 27 million gallons of alcohol monthly for making synthetic rubber.

Closely related to ethyl alcohol is methyl alcohol. Both are colourless liquids, but methyl alcohol is considerably more volatile. Formerly it was largely obtained by heating wood in retorts and condensing the products driven off; for this reason it is still sometimes known as wood spirit. Although still made – to the extent of 25,000 tons annually – by this process in timber-producing countries, most methyl alcohol is now made synthetically by a process first developed by the I. G. Farbenindustrie in 1924 and now worked throughout the world. This process, carried out at high temperatures and high pressures, involves combination between hydrogen and carbon monoxide; both these gases are derived from water-gas – already described in connexion with the Haber-Bosch process – made by passing steam through glowing coke. Combination of the two gases takes place in the presence of a catalyst such as zinc oxide or copper. World production by this process is now of the order of 300,000 tons annually.

Methyl alcohol finds many uses. It is used as such in fuels for internal combustion engines – especially for racing purposes – in 'denaturing' ethyl alcohol, and as an anti-freeze in motor-car radiators. It is also an important starting-material for other chemicals, of which formaldehyde is the most important; this substance is used primarily for making certain important types of plastics, dyes, and drugs.

A modification of the process used for making synthetic methyl alcohol yields other related alcohols, including ethyl alcohol, propyl alcohol, and butyl alcohol. The first of these we have already dealt with in detail; propyl and butyl alcohols are primarily useful as solvents.

Most people know that if wine and similar liquors are left exposed to the air they become sour and undrinkable. This process is used industrially for making some types of vinegar; its basis is the fact that certain micro-organisms can convert alcohol into acetic acid. Much synthetic acetic acid is also made, as we have seen, by various transformations from acetylene. Very large quantities are used for making cellulose acetate, an important synthetic textile, and amyl acetate, a useful solvent, and for other purposes.

Certain other organic acids normally associated with plants are also of industrial importance. In the fermentation of wine, for example, an insoluble deposit forms in the vats. This is a very impure form of tartaric acid; it is purified by treatment with sulphuric acid. Tartaric acid is used as a mordant, in making some soft drinks, and for the manufacture of baking-powder. Citric acid, originally obtained from citrus fruits such as lemons, is now made on a large scale by a fermentation process. Most of it is used for the making of lemonade; smaller quantities in making blueprints and for other photographic purposes, in analytical chemistry, and in pharmacy. Oxalic acid, a poisonous constituent of rhubarb and other leaves, is made synthetically, in the form of its sodium salt, by reaction between carbon monoxide and caustic soda; the sodium salt of formic acid is an important intermediate product. Although not produced in great quantities oxalic acid finds many uses; for example, in making metal polishes, in purifying waxes, in making dyestuffs, in bleaching straw, and as an analytical chemical. Formic acid, a pungent and caustic substance present in the fluid ants inject on biting, also finds many uses – in silvering glass, for coagulating raw rubber, in electroplating, and in the making of other chemicals. Lactic acid – the acid of sour milk – is made industrially by fermenting sugary and starchy waste products similar to those used for making ethyl alcohol. It is used in tanning, in certain pharmaceutical products, and as a substitute for citric acid in making certain soft drinks.

The making of many organic chemicals, especially dyes and drugs, requires many intermediates which are scarcely used at all, or even known, outside the industry. A great number of these are made by carrying out standard chemical processes on primary raw materials such as benzene, toluene, naphthalene, phenol, and so on. In order to sort out these very complicated processes without going into too much purely chemical detail, it is convenient to discuss each of these basic processes in general terms.

The simplest way of ringing the changes on coal-tar derivatives is by oxidation, but unfortunately this usually leads to a mixture of products, difficult to separate, instead of only one. There are, however, some important exceptions; phthalic anhydride, for example, is simply made by passing air through hot naphthalene in the presence of a catalyst. Toluene can be oxidized to benzaldehyde and anthracene to anthraquinone.

Treatment with nitric acid, mixed with oleum, brings about the important process of nitration, a means of introducing the element nitrogen into the carbon-hydrogen structure of coal-tar hydrocarbons. A typical product is nitrobenzene, a yellow oily liquid easily converted to the extremely important substance aniline. Toluene can similarly be converted into nitrotoluene and thence to toluidine.

Treatment of the coal-tar hydrocarbons and their relatives with oleum alone converts them into the corresponding sulphonic acids. These are capable of many further chemical transformations; for example, if fused with caustic soda they are converted into phenols, substances related to carbolic acid. Thus naphthalene can be converted into naphthol; on being treated with oleum, naphthol is converted into various naphthol-sulphonic acids which, because of their great reactivity, are among the most important of all dyestuffs intermediaries.

When the coal-tar hydrocarbons and their relatives are treated with chlorine, under suitable conditions, the chlorine is either added on or some hydrogen is driven out and replaced by chlorine. Some of these derivatives are important in themselves; a form of benzene hexachloride, for example, is the exceedingly valuable insecticide Gammexane. More important, however, is the fact that these chlorine derivatives can be used as stepping-stones to many other products.

The new and unfamiliar chemical names which were of necessity mentioned in the paragraphs above – anthraquinone, benzaldehyde, toluidine, naphthol, and so on – are not important in themselves. What it is important to realize is that they form essential stepping-stones to the immense range of dyes, drugs, insecticides, and other substances which are the ultimate products of the organic chemical industry and that they are derived from comparatively simple and familiar chemicals such as benzene, sulphuric acid, and caustic soda. The making of these heavy organic chemicals is in a sense an industry within an industry, for many of them are no sooner made than they are absorbed again for further chemical transformation into final products.

Among the newest products of the organic chemical industry are the detergents. Soap-making is one of the oldest of all chemical industries, but until only a few years ago soap had no rival. To-day, however, as a result of world shortages of both fats and alkali and of certain disadvantages of soaps in some special industrial applications, the position is altogether different. Many substances have been developed which have properties similar to soap but are of an entirely different chemical nature. In the last few years their manufacture has grown very rapidly. In the United States, for example, detergents are now made to the extent of 350,000 tons a year, roughly equivalent to one-quarter of the normal soap production. In Britain, too, detergents are now in use in almost every household, as well as finding many important industrial applications. The normal raw materials for making detergents are petroleum or coal-tar hydrocarbons; for some detergents, however, the raw material is fat.

Detergents are widely used for washing clothes both

in the home and in laundries. They have the advantage of lathering equally well in hard and soft water; in hard water they do not form the insoluble scum which soap does. They are used also for removing dirt and grease from raw wool; as wetting agents to ensure that insecticide sprays penetrate thoroughly; in shampoos and other toilet preparations; to promote evenness of colour in dyeing; for degreasing metals; and for many other purposes.

A particularly interesting branch of the organic chemical industry is that concerned with the preparation of perfumes and flavouring essences. Many of these are derived from natural sources such as flowers and plants. For example, citral, citronellal, and cinnamaldehyde are aromatic substances extracted from oils of lemon grass, citronella, and cassia respectively. From oil of caraway is derived a substance known as carvone, which is used in making kümmel and other liqueurs. Many of these natural products are now made synthetically; in addition the organic chemist can now offer a wide range of pleasantly aromatic substances which are unknown in nature. This branch of the industry may be said to have had its beginnings in Britain, for as long ago as 1868 Perkin synthesized coumarin, a valuable natural aromatic substance largely responsible for the odour of newly mown hay. Natural coumarin is obtained from the tonka bean; it is used for scenting tobacco, soap, and cosmetics. A number of artificial musks are made by treating a coal-tar hydrocarbon with nitric acid. A substance known as terpineol, a lilac-like perfume much used in the soap industry, is made from pinene, contained in oil of turpentine, by a series of complicated chemical transformations. Vanillin, which gives natural vanilla its characteristic flavour, is made in considerable quantities from eugenol, which comprises almost the whole of oil of cloves, and from waste products of the manufacture of wood pulp. From oil of anise is obtained a substance called anethole; this can be transformed into anisaldehyde, which has an odour of hawthorn.

A substance with which many have for many years become rather too familiar may appropriately be mentioned at this point. This is the synthetic sweetening agent saccharin, accidentally discovered in the United States in 1879 by the chemists Remsen and Fahlberg. This substance is 550 times sweeter than cane-sugar and when the latter is scarce, as in time of war, is a valuable alternative to it. It is also valuable to diabetics and others requiring a sweetening agent which is not a carbohydrate. Another sweetening agent – dulcin – was discovered in 1892; chemically this is p-ethoxyphenyl urea. It has a very pure taste and is free from the after-taste which some people find objectionable in saccharin. At the end of World War II Dutch chemists added another very important substance to the list of artificial sweetening agents. This is an orange crystalline substance known as propoxynitraniline; it is more than four thousand times sweeter than cane-sugar. Yet another artificial sweetening agent, for which absence of after-taste is also claimed, is Sucaryl.

With the growth of the organic chemical industry, coal – of supreme importance in Britain since the Industrial Revolution – has assumed a new role which it is important to understand. As has been explained, organic chemicals have the common property of containing the element carbon. Most organic compounds consist very largely of carbon; phenol (carbolic acid), for example, contains 75 per cent by weight of carbon.

For manufacturing organic chemicals, therefore, it is
essential to have a raw material rich in carbon. In practice this means either coal, containing about 75 per cent
by weight of carbon, or petroleum, which contains
rather more. In Britain, and other countries possessing
coal but no oil, coal is, as one might expect, the main
source of carbon for the organic chemical industry.

Coal itself is a substance so complex that in most
instances the carbon must be transformed before use
can be made of it. An important source of usable carbon
is coal-tar, a by-product of the making of coal-gas. The
use of this dates back, as we have seen, to the time of
Perkin. In addition large quantities of coal are coked in
special coke-ovens also yielding tar. From the coke,
itself rich in carbon, are made organic chemicals such
as methyl alcohol and calcium carbide. On purification,
accomplished chiefly by distillation, coal-tar yields a
variety of products. The most important is benzene,
which can be turned into other products such as nitro-
benzene and anilin. Coal-tar also contains about 10 per
cent by weight of a white solid called naphthalene, an
extremely important chemical intermediary especially
valuable for making dyestuffs. Closely related to naphthalene is anthracene, another coal-tar constituent.
Another valuable tar chemical is toluene, used in
making a range of products from TNT to saccharin.

The use of petroleum as a raw material resembles
that of coal, though the chemical manipulation is
necessarily rather different. Simple distillation yields
simple compounds, such as benzene and toluene, which
can be used directly by industry; the higher fractions
can be 'cracked' – or decomposed by strong heat – to
form the simpler substances most in demand. As might
be expected from the existence of immense deposits of

petroleum within its own frontiers the lead in the use of petroleum as a chemical raw material has been taken by the United States, which now uses some 2 million tons annually for this purpose.

In Britain the petrochemical industry has, chiefly for economic reasons, grown more slowly, but the erection of large oil refineries during and since the war has given the chemical industry important new sources of raw materials, especially ethylene and propylene. Five 'cracking' plants, aimed primarily at ethylene production, have been built; as a result consumption of ethylene has increased tenfold since 1950 – now standing at some 70,000 tons annually – and by 1960 it may be twice this. Xylene, necessary for the manufacture of Terylene, is being made from petroleum.

The use of both coal and oil as a raw material for the chemical industry presents a common difficulty. Natural deposits of both are limited in extent and irreplaceable; both are also needed in immense quantities as a fuel. Strict economy in their use is therefore urgently necessary.

CHAPTER TWELVE

Medicine and the Chemical Industry

THE products of the chemical industry which are used in medicine come almost entirely under the heading of fine chemicals, by which is meant substances produced in small quantities in a high state of purity. Fine chemicals are also generally taken to include highly purified forms of substances – such as paraffin oil and sodium sulphate – which are met with in a cruder form as heavy chemicals.

Among the earliest medical products of the chemical industry are disinfectants. These came into use well before the discovery of the bacterial nature of many diseases which gave a logical explanation of their action. For general sanitary purposes it is necessary to have chemicals which are available cheaply in considerable quantities, which are effective in destroying bacteria, and comparatively safe and easy to handle. Bleaching powder, or chloride of lime, fulfils these various conditions admirably and is widely used, for example, for military sanitation in the field. Coal-tar provides another valuable source of disinfectants for general purposes, for certain fractions of it are rich in carbolic acid – Lister's original disinfectant – and related substances. Of these related substances cresol is particularly important; it is not very soluble in water but can be emulsified with the aid of soap. Such powerful but crude disinfectants are not, however, suitable for purposes such as the sterilization of surgical instruments, the making of mouth-washes, cleansing wounds, and so on. For these purposes a number of fine chemicals are made. Some, such as hydrogen peroxide, we have already met; substances of this class depend for their effect on their powerful oxidizing action. Potassium permanganate, distinguishable by the deep purple colour it gives in solution, is a substance of this type; it is painfully familiar to most people as an ingredient of the pink mouth-washes used by dentists. Potassium chlorate, another oxidizing agent, is similarly used in mouth-washes, throat pastilles, and so on. Iodine, one of the earliest antiseptics, is still widely used, though perhaps not as widely as its remarkable properties deserve. It is a powerful germicide; moreover, it has great powers of penetration. This last is a factor whose supreme importance is not at all generally appreciated. There are many substances which rank as excellent disinfectants when tested under ideal conditions in laboratory cultures. In practice, however, they may prove disappointing because they completely fail to penetrate to microbes lurking in particles of wound debris, pus, and other strongholds. Consequently, the infection breaks out afresh as soon as the antiseptic has dispersed.

In recent years other members of the family which includes carbolic acid and cresol have been investigated in the hope of finding new substances possessing their powerful germicidal properties, but lacking their destructive effect on human tissues. This search has met with considerable success, for example in making available antiseptics of the 'Dettol' type; this is essentially a chlorine derivative of a phenol known as *m*-xylenol and has proved particularly useful in obstetrics. Other comparatively new but already important antiseptics of this type are amyl-*m*-cresol and hexylresorcinol.

Before passing on to chemotherapy – a field in which chemistry and medicine have been linked with immense success – some attention must be paid to nomenclature. Hitherto we have spoken of antiseptics, disinfectants, and so on without specifying exactly what these terms imply, because they are familiar and in their previous contexts exact definitions were not important.

A disinfectant is a germ-killing substance – a germicide. An antiseptic is a substance which prevents bacteria multiplying – a bacteriostatic. Some substances perform both functions; others are antiseptics at low concentrations but disinfectants in strong solutions. As a class, disinfectants and antiseptics are general poisons attacking all living cells, whether they be those forming parts of animals or individual bacteria; they may, however, be fairly selective in their action and in general the most satisfactory antiseptics and disinfectants are those which are more poisonous to bacteria than to animal cells. When the selectivity is comparatively high the substances can be used for treating wounds, boils, and other local infections.

Bacterial infection is, however, most dangerous when it affects the body as a whole, as in infectious diseases such as typhoid, cholera, and syphilis, and when infection has spread beyond some local focus such as a septic wound. In these circumstances the kind of substance we have so far considered is useless. If enough is introduced into the blood stream to destroy the invading bacteria the tissues of the body are themselves severely poisoned. Even if this does not occur, the body tissues may so rapidly absorb the drug that none is left for attacking infection.

It is not surprising therefore that ever since the role of bacteria in disease was realized both medical men

and chemists have dreamed of a substance which differed so greatly in its toxicity towards animal cells and
to the bacteria which cause disease that it could be used
within the human body. Ideally, they sought a substance deadly to bacteria but completely harmless to
human beings. This ideal has now been partially
realized in penicillin; partially in the sense that although
innocuous to human beings it is not active against all
the bacteria of disease, but only against a considerable
range of them.

To speak of penicillin is, however, premature, for
this is the culmination of more than half a century of
intensive – and for a long time unrewarded – research.
Although Paracelsus – centuries before the existence of
bacteria was suspected – had used chemicals in attempting to treat what we now recognize as bacterial infections the story really begins in the year 1880. By then
Lister and Pasteur had, by their independent researches,
demonstrated the immense practical importance of
antiseptics and hopes were high that it might be possible to find a drug capable of treating general bacterial
infections – what we now call a chemotherapeutic
agent.

Robert Koch, the great German bacteriologist, pursued interesting researches with mercuric chloride; he
found that this substance is deadly to anthrax bacilli but
apparently so much less poisonous to animal cells as to
hold out the hope that it might be used for treating
anthrax. All his experiments – made with guinea-pigs –
were, however, failures; we now know that one reason
for this is that proteins in the blood absorb mercuric
chloride so rapidly that it never comes into contact
with the bacteria it is intended to attack.

Disappointment followed disappointment – in ten

years of intensive search not a single substance was found capable of effectively attacking a single species of bacteria within the body. It seemed, indeed, that there must be some natural law governing the relative toxicity of all substances to bacterial and animal cells which made the search hopeless from the outset. In 1890 Behring, discoverer of the diphtheria anti-toxin, wrote dispiritedly that 'the pessimism of him who declared disinfection in the living body to be for all time impossible appears to be only too justified'. Fortunately for mankind this pessimism proved unjustified.

In 1885 Paul Ehrlich, another great German scientist, showed that a common synthetic dye – methylene blue – was selectively absorbed by the malaria parasite, the cells of the human body being almost untouched, but unfortunately he soon abandoned this chemical line of research and devoted himself to the biological control of disease by means of anti-toxins. In 1904, however, he returned to his chemical work with a Japanese assistant, Shiga. Within five years this rather curiously assorted partnership proved very fruitful. It produced the valuable arsenical drug salvarsan. As early as 1900 it had been shown that an arsenical compound named atoxyl – first made as long ago as 1863 – was useful in treating certain skin diseases and in 1905 it was found to be useful against the organisms, trypanosomes, which cause sleeping sickness in human beings and ngana in cattle. Atoxyl was, however, a dangerously poisonous drug to use. By modifying the chemical structure of atoxyl Ehrlich first produced arsacetin and ultimately his world-famous anti-syphilitic salvarsan or '606' – so-called because it was the 606th derivative of atoxyl that he and Shiga tested. Ehrlich's achievement was recognized by the award of a Nobel

Prize in 1909 and its greatness may be measured by the fact that not for more than thirty years was any new drug available which improved upon salvarsan and its various modifications for treating syphilis. With salvarsan was born the now immensely important chemotherapeutic branch of the chemical industry.

Although Ehrlich's original purpose was the discovery of a drug for treating trypanosome infections he ended up, as we have seen, with a specific against syphilis. The success of salvarsan, which effectively refuted the widespread pessimism about ever finding useful chemotherapeutic agents, stimulated new research on an agent for attacking trypanosomes, which in tropical countries exact a heavy toll of the lives of both human beings and cattle. About 1920 the German firm of Bayer launched a drug named Germanin on the market for this purpose; French chemists discovered its identity and marketed it as Fourneau 309. It is colourless, but chemically not dissimilar from a dye named Trypan Red which Ehrlich had synthesized in 1904. In 1919 chemists of the Rockefeller Institute discovered a powerful and relatively non-toxic arsenical drug for attacking trypanosomes; to this was given the name tryparsamide.

In 1932, the whole picture of chemotherapy was dramatically changed. Domagk, working in the Bayer laboratories, noted that a red dye, named prontosil, would destroy certain types of streptococci, among the most dangerous of infective bacteria, but yet was almost non-toxic to the cells of the human body. Not until 1935, however, was this discovery announced to the world. Then French chemists at the Pasteur Institute in Paris discovered that only a part of the prontosil molecule was active; equally good effects could be obtained by means of a very much simpler substance known as sulphanilamide. Exhaustive tests at Queen Charlotte's Hospital in London confirmed this astonishing result. Sulphanilamide, a well-known and simple substance long used as an intermediary in making dye-stuffs, proved a far more useful chemotherapeutic agent than all the complex derivatives with which chemists had struggled for years previously. The effect was dramatic because the simplicity and cheapness of the drug made its use on the widest scale throughout the world immediately possible. It proved highly successful in treating such diverse diseases as puerperal fever, septicaemia, gonorrhea, pneumonia, and certain types of meningitis.

Chemists were not content with sulphanilamide itself. By 1942 no less than 3,600 related substances – collectively known as sulphonamides – had been made and tested. Of these a few proved to have definite advantages for certain purposes and came into general use. The first of these was the world-famous M and B 693, a British contribution by the firm of May and Baker. Others are sulphadiazine, sulphaguanidine, sulphasuxidine, Sulphamezathine, and sulphathiazole.

Meanwhile, following the success of the arsenical drugs already mentioned, the special problems of tropical medicine had received much attention. In 1935 an important new anti-trypanosome – synthaline – was introduced and was followed by a succession of derivatives. In Germany the very active firm of Bayer produced the synthetic drug Mepacrine (Atebrin) as a valuable substitute for the quinine which for three centuries had been the only known specific against malaria. This was manufactured by the Allies in huge quantities during World War II, and in the highly malarious Pacific theatre proved immensely important after the loss of Java, the world centre of quinine production. A measure of the significance of malaria is given by the fact that conservative estimates place the total number of sufferers at any one time as high as 700 million, or one-third of the world's population; the number of deaths probably exceeds 3 million a year. In 1943 another notable advance in this field was made by chemists of Imperial Chemical Industries Limited, who produced a totally new kind of anti-malarial known as Paludrine. This is a very powerful drug, which has proved itself in extensive trials, and has the advantage over both quinine and Mepacrine of attacking the malaria parasite at almost all stages in its complex life-cycle instead of in only a part of it.

The discovery and development of penicillin is not merely the greatest achievement in chemotherapy, but also one of the greatest achievements in the whole history of medicine. Penicillin was discovered in 1928 by Professor (now Sir Alexander) Fleming, who noted that a mould which had by chance grown on a culture of staphylococci was devouring the bacterial colonies. Investigation proved that the bacteria were being destroyed by a substance produced by the mould, but preliminary chemical experiments indicated that it was so sensitive and easily destroyed that its purification would be a matter of very great difficulty. Furthermore, there was at that time no indication that this particular antagonism between mould and bacteria was caused by so unique a substance as penicillin ultimately proved to be. Many apparently similar examples were already known and had been closely investigated, but when the active substances had laboriously been isolated they proved unsuitable for use in medicine. By 1932 interest in penicillin had almost completely lapsed as a result of the great technical difficulties, and the great possibilities of new synthetic drugs, and it remained dormant until 1939. In that year Professor (now Sir Howard) Florey and Dr E. Chain, of the University of Oxford, became very much interested in the general problem of antagonism between moulds and bacteria and they decided to undertake, with a team of specialists, a systematic investigation of all examples of the phenomenon then known. By a fortunate chance penicillin ranked high among the examples chosen for investigation, but it was chosen much more for its scientific interest than in the expectation that any valuable chemotherapeutic agent would result.

The difficulties were great, but by May 1940 enough of a solid preparation of penicillin had been laboriously got together for the treatment of experimental bacterial infections in mice. The results were dramatic even though this first material consisted, as we now know, of no less than 99 per cent impurity. By February 1941 sufficient material was available for the first test on a human patient; it failed because the quantity of the drug was insufficient for treatment to be completed. Nevertheless success was so nearly achieved that the latent possibilities of penicillin became apparent.

At once plans were made for the preparation of penicillin on a much larger scale and the co-operation of the British chemical industry was sought and obtained, despite the immense difficulties resulting from wartime conditions. Trials of the much larger quantities of material thus available clearly revealed the truly remarkable properties of penicillin. Not only would it do all and more than the sulphonamides would, but it was completely without ill effect on the patient. At long last science had found what Ehrlich called a 'magic bullet' – a substance which would seek out and destroy invading bacteria but leave the human patient unaffected.

Professor Florey and those most intimately concerned in the British project realized that in view of penicillin's remarkable value in treating infected wounds it could be, if available in quantity, of immense value in the war, for the infection of wounds is a principal cause of death or permanent disablement among military casualties. Even in normal times the problems of extracting pure pencillin from the broth upon which the mould had grown was one which would tax the resources of the chemical industry to the limit, for it must be remembered that less than ten years previously this had proved an insoluble problem even on a miniature scale within the walls of well-equipped laboratories. For the British chemical industry, shouldering many new wartime responsibilities and sorely hampered by frequent heavy air-raids, the establishment on a large scale of plant requiring the previous solution of many major problems of a kind never before encountered proved virtually impossible. Nevertheless a small penicillin factory was established at Manchester and provided material – at a critical moment – for far more extensive trials than had previously been possible.

Accordingly, in the summer of 1941 Professor Florey and Dr Norman Heatley, a member of his research team at Oxford, visited the United States and in an extended tour sought the assistance of American scientists and chemical industrialists. All the information so laboriously collected in England was made available without reservation. After much hesitation – for the claims made and subsequently justified seemed at that time fantastic – the American chemical industry was won over. Once convinced, however, co-operation was on a scale commensurate with the prize to be won. After intensive preliminary research – of which the most outstanding result, of decisive importance, was the discovery of a way of making the mould produce several times more penicillin than normal – several great factories were established in various parts of the United States.

So rapid was industrial progress that by D-Day, 1944, sufficient penicillin was available to treat all casualties in the Allied forces; it had also previously been extensively used in many other theatres of war. The number of human lives saved in this way alone must be numbered in tens of thousands. Since the war American production has been very much increased and in Britain, freed from the insuperable wartime difficulties, a number of very large factories have been established. In Sweden, Italy, France, and many other countries penicillin factories of varying size have been set up. The capital invested must be measured in tens of millions of pounds. In consequence penicillin is to-day available in most parts of the world in quantities sufficient to meet all needs.

As measured by both the technical difficulties to be overcome and the benefit to mankind the manufacture of penicillin ranks as one of the most outstanding achievements in the whole history of the chemical industry. Most significantly, too, this achievement was made in a field in which Germany was accustomed to lead the world.

Valuable though it is, penicillin is by no means a universal panacea for the treatment of bacterial infections, for it has certain well defined limitations. There are, for example, certain types of infections, such as those of cholera and tuberculosis, which are unaffected by it. Again, certain strains of common bacteria which are normally very sensitive to penicillin may have a natural resistance to the action of the drug or may acquire such resistance during treatment. Such resistance is, however, almost always specific to a particular drug; bacteria which are not sensitive to sulphonamides respond to penicillin and vice versa.

It is scarcely surprising that in an attempt to fill these gaps in the chemotherapeutic armoury attention has been focused during the last ten years on other antibacterial mould products, or antibiotics as they are now called. Literally thousands of examples of antagonism between bacteria and other micro-organisms have been closely studied and the responsible substances isolated and studied in detail. Results have been rewarding, though no new antibiotic has been found which is of the same order of importance as penicillin itself. The introduction of the latter has set completely new standards in medicine. In describing a new antibiotic, cephalosporin, in 1951, Sir Howard Florey and his collaborators wrote: 'It is perhaps a measure of progress in the treatment of bacterial disease that little more than ten years ago cephalosporin P_1 would have been considered a substance of outstanding qualities'. Today, however, the practical value of cephalosporin is treated with great reserve.

A number of new antibiotics are, however, in regular industrial production by methods generally similar to those for penicillin. The best known of these is streptomycin, a promising agent in the treatment of tuberculosis, certain types of meningitis, and other infections. Others are chloromycetin, aureomycin, and terramycin, which all find extensive use for special purposes. Some are made synthetically.

The success of the antibiotics, whose full value is certainly still far from exploited, has temporarily directed attention away from experiments aiming at synthesizing new chemotherapeutic agents rather than utilizing those which nature provides. Nevertheless synthesis is still producing results and there can be no doubt that many important new drugs will yet result from this source. Already, for example, a class of drugs known as the sulphones has provided, for the first time, a hope of getting to grips with one of the most feared diseases – leprosy. Other compounds – of hydrazine and nicotinic acid – have shown some promise for the treatment of tuberculosis.

Although so much has been achieved within the comparatively brief space of twenty-five years one can say with confidence that the chemists' contribution to the conquest of disease has only just begun. In this conquest the chemical industry plays a vital role for, apart from discoveries – such as that of Paludrine – made wholly within its own research organizations, it alone can produce on a large scale the drugs whose practical value has been demonstrated in the laboratory.

From this survey of the principal achievements in chemotherapy we turn to another field of joint chemical-medical enterprise in which no less valuable results have been achieved – the field of anaesthetics.

In former days the surgeon had a double difficulty to overcome. On the one hand the mere physical shock of operation might prove fatal to his patient and in any event the need for speed was so paramount as to exclude the delicate dissection which alone might save life. On the other hand, wound sepsis was the rule rather than the exception after operation, and patients who survived the operation itself more often than not died later from bacterial infection. Chemotherapy, as we have just seen, has largely relieved the surgeon of the second of these risks. Anaesthetics, besides immeasurably relieving the sufferings of the patient, enable the surgeon to carry out prolonged and delicate operations, for example those involved in brain surgery, without distraction.

Britain may well take pride in always having taken an outstanding part in the development of anaesthetics, the story beginning as long ago as 1794. In that year Thomas Beddoes, founder of the Pneumatic Institution in Bristol, found that ether was valuable in relieving pains in the chest. Beddoes was followed by Humphrey Davy, who in 1799 noted that nitrous oxide gas produced a kind of intoxication in which pain is deadened and might 'probably be used with advantage during surgical operations in which no great effusion of blood takes place'. Although the hilarious intoxication produced by nitrous oxide made 'laughing-gas parties' fashionable in the early nineteenth century the medical profession failed to avail itself of the valuable new tool which lay ready to its hand and nearly fifty years passed before it was used in surgery.

Laughing-gas parties spread to America as well as what were called 'ether frolics'. In 1842 a Georgian doctor, C. W. Long, introduced ether for minor surgical operations. In 1844 a Connecticut doctor, Horace Wells, and a dental colleague named Morton, experimented with the use of nitrous oxide in dentistry, but with mixed success which excited adverse criticism. Undaunted, they turned to ether, and on 16 October 1846 they gave a convincing demonstration of the use of ether as an anaesthetic before a distinguished medical audience in Boston. This was a turning point in surgery; Sir William Osler, the great surgeon, rightly described the introduction of anaesthetics as the greatest single discovery in the history of medicine. Ether was soon accepted throughout the world. In 1847 Sir James Simpson of Edinburgh discovered the valuable anaesthetic properties of chloroform. The supply of these substances in a pure form was, of course, at once a new task for the chemical industry.

One important and now almost universally accepted use of anaesthetics originally excited much controversy; this was their use in relieving the pains of childbirth. In 1853 this controversy, which now seems almost incredible, was killed when John Snow administered chloroform to Queen Victoria at the birth of Prince Leopold. Although chloroform proved more dangerous than ether the greater control possible in its administration soon made it the more popular of the two.

The original anaesthetics were all what are now called inhalation anaesthetics. For a long time ether, chloroform, nitrous oxide, and to a lesser extent ethyl chloride, ruled supreme. In recent years this 'big four' has been joined by others of which two are outstanding. One of these is a simple hydrocarbon gas called cyclopropane; the other is a chlorine compound – trichlorethylene or Trilene – which we have already referred to as a valuable solvent and dry-cleaning agent. For use as an anaesthetic, however, trichlorethylene must be very much more carefully purified than the ordinary product. It has a distinct advantage over ether and cyclopropane in not being inflammable.

The medical use of local anaesthetics – substances which deaden pain at the immediate site of operation as opposed to producing complete unconsciousness – dates from 1884, when Koller introduced cocaine for ophthalmic surgery. In this he was in a sense putting ancient practice on a firm footing, for from the earliest days physicians had tried to eliminate the worst horrors of surgery by the use of natural pain-killers such as opium and mandragora. To the great physician Sydenham is attributed the aphorism that 'few would be willing to practise medicine without opium'. Morphine, the active principle of opium, is to-day an important product of the chemical industry. In early days, however, so little was known about the action of these analgesic drugs that their use was very hazardous. Koller revived interest in local anaesthetics at a time when medical and scientific knowledge had advanced sufficiently to allow their use with comparative safety.

Procaine or novocaine, a synthetic chemical related to cocaine, was introduced in 1905. Other synthetic drugs followed, such as nupercaine and amethocaine; one object sought was a means of prolonging the action of the drug beyond the forty minutes or so of cocaine itself.

Anaesthetics administered by injection into a vein were used as long ago as 1872, when the French physician Oré used chloral hydrate for the purpose. They did not become popular, however, until 1932, when the barbiturates were introduced; an important modern example of this class of synthetic anaesthetics is Pentothal.

The manufacture of anaesthetics, both in the quantity of material produced and in the number and complexity of the products, makes important demands on the resources of the chemical industry. The benefit to the community is, however, out of all proportion to the effort required. In these days, when science is so often referred to as having become an agent of destruction, it is timely to reflect on the immense contributions to human welfare made by chemotherapy and anaesthetics alone. If a balance-sheet were drawn up, with the lives lost through the use of science in war put on one side and the lives saved by chemotherapy, anaesthetics, and other similar substances, on the other, the balance would not merely be on the latter side – it would be overwhelmingly so. For every life which has been destroyed through misapplied science perhaps a hundred – as a conservative estimate – have been conserved.

Although chemotherapeutic agents and anaesthetics are the two outstanding contributions of the chemical industry to modern science they are very far from being the only ones. The challenge offered by the discovery of vitamins and hormones has been accepted with alacrity. As is generally known, the vitamins are food factors which play a dominant role in human health. Deficiency of vitamin C, for example, causes scurvy, once the dread of all deep-sea sailors. Lack of vitamin D causes rickets. Deficiency of vitamin B_1, prevalent among Eastern populations subsisting on polished rice, causes beri-beri. Lack of B_2 causes pellagra. The supplying of the essential vitamins either to enrich natural foods such as bread or margarine or for making pharmaceutical concentrates is an important task of the chemical industry, which fulfils this function in two ways. On the one hand it manufactures pure vitamins, or very concentrated forms of them, from natural materials. On the other, it makes certain important vitamins synthetically. As almost all the vitamins have very complex chemical structures these syntheses on an industrial scale are outstanding achievements.

Of the synthetic vitamins, vitamin C is perhaps the most important. Although the laboratory synthesis was first achieved in Britain – it was, indeed, the first vitamin ever to be synthesized – the first industrial synthesis was effected in the United States in 1934. To-day immense quantities are made, annual output being about eight hundred tons. This figure should be set against the average adult requirement, which is of the order of one-third of an ounce per year. Another extremely important synthetic vitamin is vitamin B_1, annual production of which is also measured in hundreds of tons; the process used involves seven distinct steps, apart from those needed to derive the starting materials.

The vitamin known as B_2 is a complex one, with at least three constituents – riboflavin, nicotinamide, and folic acid. All three have been synthesized. Synthetic riboflavin and nicotinamide are made on a very large scale. The enrichment of white flour, from which rather paradoxically natural vitamin B_2 has been removed by milling, is an important use for the synthetic product. In the United States, for example, some 40 million barrels of flour are enriched annually, requiring 16 tons of riboflavin and 130 tons of nicotinamide. Other vitamins of the B group which are manufactured synthetically are pyridoxin (B_6), possibly not essential to man though essential to certain animals, and B_{12}; the latter, a cobalt-containing substance, has proved identical with a factor present in liver extracts which prevents pernicious anaemia.

Another vitamin available synthetically is vitamin E, believed to play some role in female fertility; its medical importance is, however, still obscure. Still another synthetic product is vitamin K, which plays an important part in controlling the clotting of blood; for this reason it finds important medical applications in controlling certain types of haemorrhage. It has also been used in the treatment of chilblains.

Allied, in a sense, to the vitamins, are the hormones. These are complex organic substances which, although utilized in only very tiny quantities, like the vitamins, exercise a decisive role in the metabolism of the body as a whole. Unlike the vitamins, however, the hormones are manufactured within certain glands of the body and are not absorbed from food.

The simplest of the hormones is adrenaline. Excessive production is associated with anger and fright and it has a marked power of contracting the blood vessels; it is often used for stopping the bleeding of small cuts. In the body it plays an important role in the transmission of nervous effects. Adrenaline is relatively easy to synthesize, the starting material being catechol, and it is manufactured on a large scale.

One of the most important of the hormones is insulin, for it is a specific for the treatment of diabetes, a very serious and common disease in which the body is unable to metabolize sugar normally. With the help of insulin, however, diabetics can be kept in good health and lead normal lives. The drug has yet to be synthesized and the very large quantities required are extracted from the pancreas ('sweetbread') of animals such as cattle, pigs, and sheep. Insulin was first isolated in Canada, as recently as 1922, by Banting and Best.

Another hormone of great medical importance is thyroxine, principal product of the thyroid gland. This can now be made synthetically, the first synthesis being achieved in Britain. Disorders of the thyroid are associated with general disturbances in the body's metabolism, often outwardly reflected as goitre.

Another group of hormones, which chemically are known as steroids, are of the utmost importance in relation to the sexual functions. The total synthesis of these substances presents immense difficulties and only limited success has been achieved even within the laboratory, though their interconversion is often relatively simple. Certain complex synthetic substances have, however, been shown to possess activities similar to those of the natural sex hormones and these synthetic substances, among which the British discovery stilboestrol is outstanding, are of the greatest importance in medicine, and are manufactured on a large scale.

Very recently yet another hormone, cortisone, has enjoyed widespread publicity, primarily because of its dramatic effects in the treatment of arthritis. Its full importance remains to be assessed, though it is already clear that the possibilities are great. The drug is at present very costly and difficult to manufacture as the main starting material is a minor constituent of ox-bile; substances contained in an inedible yam found in Mexico are, however, possible alternatives as starting materials. It has been found that a substance known as ACTH, present in the pituitary glands of hogs, has the power of stimulating cortisone production by the adrenal glands and is therefore an effective substitute for cortisone itself. The supply of ACTH, however, still provides great difficulties as the pituitary glands of nearly half a million hogs are required to prepare one pound of the drug. Despite these difficulties the chemical industry, notably in the United States, is already producing considerable quantities of both cortisone and ACTH.

The list of the contributions of chemistry to medicine

– translated into practice by the chemical industry – is almost endless. There are, for example, the antipyretics or fever-reducing drugs. Up to 1885 pride of place in this group was held by quinine, but to-day various synthetic products such as aspirin, phenacetin, and antipyrine are very widely used. Closely allied to aspirin, in the chemical sense, is methyl salicylate or artificial oil of wintergreen; this is used in embrocations. Aspirin and phenacetin are commonly combined in modern antineuralgics with caffeine and codeine.

The use of herbal preparations is traditional in medicine, but in former times ignorance of the real functions of vegetable drugs and the great variations in the biological activities of different preparations made their effects unpredictable and often hazardous. To-day the picture is entirely different; in many cases it is possible to extract the active principles of medicinal plants, so that they may be dispensed unequivocally by weight, or alternatively to standardize carefully prepared products so that dosage can be accurately controlled. The number of vegetable drugs in regular use is legion. Apart from quinine and morphine, already described, mention may be made of strychnine, hyoscine, digitalis, ergot, atropine, aconitine, ephedrine, and physostigmine. In addition to the natural products industrial chemists provide a host of synthetic drugs which have similar, and sometimes more satisfactory, physiological effects.

Closely allied to medicinal fine chemicals are the veterinary products of the chemical industry. These are becoming of rapidly increasing importance as a result of the great interest now taken in all problems of animal health. To a considerable extent the two fields overlap, medicinal products finding corresponding applications in veterinary practice. Economic questions and the possibility of carrying out treatment under primitive conditions with unskilled workers are, however, frequently of decisive importance.

A recent estimate indicates that in Britain diseases of cattle alone cause an annual loss of £(m)20; roundworms in sheep alone cost Britain £350,000 annually. The United States Bureau of Animal Industry sets that country's annual losses at 418 million dollars. These figures indicate that the problem is immense, but although the serious study of animal diseases is of comparatively recent origin encouraging progress has already been made.

Phenothiazine and hexachloroethane, for example, have proved very effective in treating certain types of worm infestations in animals and are manufactured in large quantities. New insecticides such as Gammexane and DDT are proving very valuable in cattle and sheep dips for destroying the ticks which are carriers of many animal diseases. Sulphonamides have found a wide field of application, for example in treating calf 'diphtheria', certain types of mastitis in cattle, and coccidiosis in poultry. A newly discovered British drug, Antrycide, has shown promise against the deadly ngana disease, spread by the tsetse fly, which is one of the principal factors preventing the raising of cattle in huge tracts of African grasslands. Calcium boro-gluconate is manufactured for treating milk fever in cattle and the allied lambing sickness in sheep. Arsenic preparations find various uses, for example, arsenobenzol for treating blackhead in turkeys. Interesting experiments have been carried out to stimulate milk production by the implantation of pellets of synthetic sex hormones. Vitamin preparations of various kinds are used to counteract deficiencies which can be as serious in the health of animals as in that of human beings.

Inorganic as well as organic chemicals are widely used. Various proprietary saltlicks, for example, provide not merely common salt but a number of other minerals required in small quantities to maintain animals in health. Copper sulphate is used for controlling foot-rot in sheep and iodides for relieving 'wooden-tongue' in cattle.

Great though the existing achievements are, it seems that we can confidently look to the chemical industry for the production in the future of many more medical products no less valuable than those we have already been considering. The main fields of research, though young, are still very fruitful. Eager minds are, however, constantly looking for possible new fields to conquer. As a concluding example one may mention cancer, that most dreaded scourge of all mankind. It is premature to say that the cure of cancer is within our grasp – though to-day there are thousands of people alive who have been cured by our present imperfect methods – but so much new knowledge has been won in recent years and the attack is so vigorous in all kinds of laboratories throughout the world, that optimism is not unjustified. In this battle the role of the chemist is becoming increasingly important. For example, synthetic oestrogens have been extremely successful in treating a particular form of cancer to which men are subject. Certain complex hydrocarbons, characterized by chemists, have been found capable of producing cancer; understanding of their mode of action may well point the way to understanding of the way in which cancer arises naturally. Already practical advantage has been taken of this new knowledge in virtually eliminating types of cancer associated with particular trades, such as the refining of shale oil. More recently still, an interesting example of good arising from evil has come to light. Certain of the war gases and their derivatives – so deadly in 1917–18 – have been found to produce on living cells effects similar to those produced by the X-rays and other radiations widely used for treating cancer. Pursuit of this new line of chemical research may well prove very rewarding. On this note of optimism we may appropriately close this brief and, for reason of space, necessarily incomplete survey of chemistry's contributions to medicine.

CHAPTER THIRTEEN

Dyes and Pigments

A CONCISE survey of the modern dyestuffs industry is difficult, because the dyes now in regular production number several thousands. It is also complicated by the wide variety of uses to which these products are put, since the majority of articles in everyday use are coloured in some form or another. To be of practical value a dye must satisfy the demands of the user of the coloured article as such, and, therefore, apart from the actual shade, the fastness properties of each dye are of paramount importance. The actual dyeing properties, which vary from dye to dye, are only the concern of the industries using them, and since few dyes are exactly alike in this respect and the medium to which they are applied varies so enormously it is obvious that the technicalities of their usage are exceedingly complicated.

Synthetic dyes, which have almost entirely replaced natural dyes for the textile industry, are largely manufactured from the products of the distillation of coal-tar, the principal raw materials being benzene, naphthalene, toluene, and anthracene. From these relatively simple organic compounds a wide range of complex 'intermediates' is manufactured. These are made in a number of stages; the first and simpler compounds such as aniline, nitrobenzene, nitrotoluene, etc., are known as 'primaries' and are made in very large quantities. Further manufactured compounds from the 'primaries' become increasingly difficult to describe chemically, but are all known as 'intermediates' and are described by trivial names until manufacture of the crude dye is complete. When this stage is reached the dye may require purification and a considerable amount of further processing is necessary before the commercial product is marketed in a form standardized to a known strength and in the correct physical condition to satisfy the rigorous demands of the colour-using industries.

The many dyes made by the methods exemplified above fall into various chemical classes. The majority of colour users prefer the dyes to be characterized by the method of application and by their properties rather than by their chemical constitution, and hence a simpler system of differentiation into Direct, Acid, Vat, and so on is used. There are three main methods of applying dyes to fibres, depending on whether the dye is water-soluble, water-insoluble, or is formed on the fibre from soluble intermediates.

In the first class are the direct dyes, so called because they will dye cellulosic materials – of which cotton is typical – without any pretreatment of the fibre, such as the application of a mordant. These are mainly azo compounds with sulphonic acid groups appended by treatment with sulphuric acid to give them solubility in water, and they are applied by immersing the material in a hot solution of the dye for an hour or more. The adsorption of dye by the material is greatly increased by the addition of common salt to the dyebath.

Water-soluble dyes are also used in the dyeing of wool and synthetic polyamide fibres such as nylon. The miscellaneous assortment of chemical types in this class are generally called the acid dyes, and although widely different in chemical constitution they all possess at least one hydrophilic – or water-loving – group. Dyeing is carried out in the presence of sulphuric or acetic

TYPICAL EXAMPLES OF DYE MANUFACTURE

AZO DYES

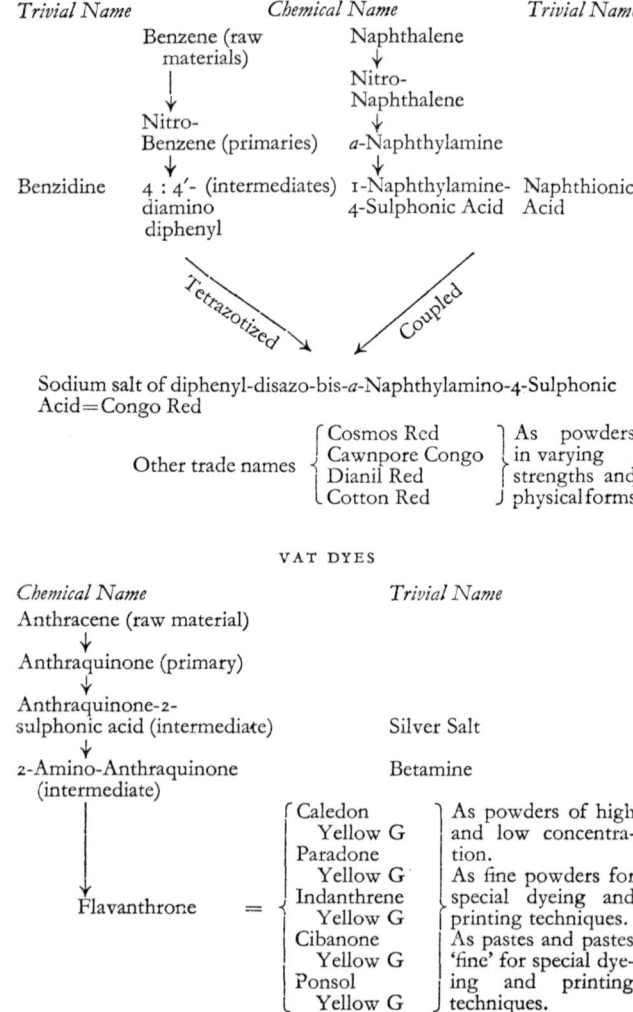

Trivial Name   *Chemical Name*   *Trivial Name*

Benzene (raw materials) → Nitro-Benzene (primaries)

Naphthalene → Nitro-Naphthalene → a-Naphthylamine

Benzidine   4 : 4'-diamino diphenyl (intermediates)   1-Naphthylamine-4-Sulphonic Acid   Naphthionic Acid

Tetrazotized   *Coupled*

Sodium salt of diphenyl-disazo-bis-a-Naphthylamino-4-Sulphonic Acid = Congo Red

Other trade names
{ Cosmos Red
Cawnpore Congo
Dianil Red
Cotton Red }
As powders in varying strengths and physical forms

VAT DYES

Chemical Name   *Trivial Name*

Anthracene (raw material)
↓
Anthraquinone (primary)
↓
Anthraquinone-2-sulphonic acid (intermediate)   Silver Salt
↓
2-Amino-Anthraquinone (intermediate)   Betamine
↓
Flavanthrone =
{ Caledon Yellow G
Paradone Yellow G
Indanthrene Yellow G
Cibanone Yellow G
Ponsol Yellow G }
As powders of high and low concentration. As fine powders for special dyeing and printing techniques. As pastes and pastes 'fine' for special dyeing and printing techniques.

acid, and during the process the temperature may be raised from tepid to boiling.

Certain acid wool dyes having certain chemical groups adjacent to the azo linkage have the power of linking up with chromium. The resultant colours are much faster to light and washing than those in which chromium is not present. They are extensively used in the dyeing of woollens and worsteds, the dye being applied simultaneously with chromium salts or treated with them afterwards.

The principal class of water-insoluble dyes is the vat dyes. The original member of this class was indigo; a number of its derivatives are most important to textile printers. Most modern vat dyes are derivatives of anthraquinone. The method of application depends upon the fact that these compounds may be 'reduced' to a form which is soluble in caustic soda solution. Cellulosic fibres will take up this reduced or 'leuco' dye from the alkaline solution; on removal from the dye-bath and exposure to the air, re-oxidation takes place and the original, insoluble dye is reformed inside the fibre, where it is retained. Vat dyes are very fast to washing and to light.

When first introduced acetate rayon presented a difficult problem in that the dyes suitable for cotton, viscose, and wool had, in general, no affinity for this fibre. Subsequently British research chemists discovered both azo and anthraquinonoid dyes having a high affinity for acetate rayon, but they are insoluble in water. It is common practice to apply this type of dye as a fine suspension in water and it is now possible to obtain acetate rayon dyed in almost any colour with this class of dye. These particular dyes are also very suitable for colouring nylon and Terylene.

The last major group of dyes comprises those formed on the material from soluble intermediates. The substances necessary for the formation of an insoluble azo dye are applied to the fibre successively. The resultant colour appears to be mechanically held inside the material, in a manner similar to that found in the vat dyes; these colours are very resistant to repeated severe washing.

One of the principal problems facing research workers is to elucidate the mode of attachment of dyes to fibre, and why certain chemical compounds have the property of being taken up by the fibre from the dyebath. The fact that most dyes have the property of being selectively taken up by certain types of fibres emphasizes the fact that there is more than one reason for the affinity of dyes. There is limited evidence that affinity may be due to one or more of the following forces.

(a) An electrostatic interaction between positive electric charges in the fibre and negative electric charges in the dye molecule, or vice versa. For example, the acid wool dyes contain negatively charged groups which, in an acid dyebath, are attracted by and attached to positively charged points on the wool molecules.

(b) An attachment by what are called 'hydrogen bonds' in which a hydrogen atom is virtually shared between the dye and the fibre.

(c) A form of long-range interaction, due to Van der Waals' forces. These forces are intermediate between true chemical bonds between atoms and attractive forces acting at a distance in much the same way as gravity does.

It must be stressed that these forces do not operate

singly and that other factors are involved, notably the shape and dimensions of the dye molecule. For example, in the direct cotton dyes the molecule must be needle-shaped, so that it can lie along the cellulose molecular chains, it must contain groups capable of linking up with the cellulose, and these groups should be so spaced as to give the maximum amount of interaction with the reacting groups of the fibre.

From this very brief account of the theory of dyeing it is clear that, even though a great deal is still unknown, enough is known to help the dye manufacturer consciously to design chemical molecules having the desired properties rather than to proceed by hit-or-miss methods.

By far the largest proportion of dyes manufactured is used for the colouring of textiles, by either printing or dyeing. Printing is essentially the localized dyeing of fabrics. The dye is mixed with a thickening agent, such as starch or gum, and transferred to the fabric by means of a stencil, engraved block, or engraved roller carrying the desired design. The printed cloth is then dried and the colour is fixed by treatment with steam. By this means all types of designs may be produced.

In addition to colouring textiles, dyes are used also to a large extent for the colouring of paper and the dyeing of leather and furs. For these purposes the majority of colours used are similar to those used for textile purposes.

Dyes are also made in oil- or wax-soluble forms. These are required for various types of polishes, for crayons and wax candles, and for tinting motor fuels and lubricating oils. These products are also useful to a limited extent for colouring certain plastics in which pale transparent shades are required.

Carbon papers and typewriter ribbons frequently contain oil-soluble dyes or insoluble pigment colours. In copying carbons for duplicating purposes alcohol-soluble dyes are necessary; the latter are also used in certain types of lacquers and wood stains.

The dyeing of aluminium is a fairly recent development. The metal is 'anodized' to form a dye-absorbent surface; the prepared surface can be dyed without obscuring its metallic appearance. The colour is fixed by treatment in hot water or by steam.

A range of specially purified dyes are used in the food industry to produce attractive colours where cooking or processing has destroyed the natural colour.

Many dyes are of the utmost importance in medical work; it is indeed no exaggeration to say that the progress of certain branches of medicine has been dependent on the availability of a large number of dyes. The reason for this has already been referred to in a different connexion. We have seen how Ehrlich was led to his important research in chemotherapy by the observation that certain cells absorb certain dyes. This specificity is widespread and is not restricted to whole cells; the microscopic structures inside cells, the nucleus, fat globules, and so on, also have specific powers of absorbing dyes. Thus it is possible to prepare thin slices of animal or plant tissue for examination under the microscope by staining in different colours the different structures which it is desired to study. Without such staining it is very difficult to identify anything at all within the cell.

Closely allied to dyes, and in everyday language often not distinguished from them at all, are pigments. The difference is essentially one of solubility or insolubility in water or other liquids. Pigments are used as insoluble particles suspended in the liquid or medium to be coloured, such as linseed oil, for paint, or rubber.

One method of classification is to make an initial division between white and coloured pigments and then to sub-divide the latter into inorganic and organic pigments according to their chemical nature. The white pigments which are most widely used are white lead (basic lead carbonates), lithopone (a co-precipitate of zinc sulphide and barium sulphate), zinc oxide, titanium dioxide, and antimony oxide. While these differ from each other in opacity, they are all sufficiently opaque to make them very suitable for the preparation of paints of good covering power. They are all, incidentally, manufactured and not natural pigments. Other white pigments are often described as extenders, because they have a much lower opacity than those already mentioned, and these include the natural mineral barytes, barium sulphate, blanc fixe (which is precipitated barium sulphate), china clay, and whitings.

Many of the inorganic coloured pigments are long established, and have come to be known by common names such as Prussian Blue, Ultramarine, and Chrome Yellow. The naturally occurring ochres are of a yellow shade and consist of hydrated iron oxides associated with varying quantities of china clay, silica, and other minerals. Where, as sometimes happens, these are associated also with quantities of manganese, resulting in a brown pigment, they are known as siennas and umbers. Ultramarine was originally produced as a pigment from the natural mineral lapis lazuli, but nowadays is manufactured from a mixture of soda, china clay, sodium sulphate, sulphur, coke, and silica. Iron oxide pigments appear in nature in various parts of the world as red oxides. Vermilion also, consisting of mercuric sulphide, was originally derived from natural deposits but is now largely made synthetically. Red lead or minium is an oxide of lead which has long been known and still widely used. Other important manufactured inorganic pigments include the Chrome Yellow pigments, ranging in shade from primrose yellow to scarlet, Prussian Blue, and the cadmium pigments. The latter consist of cadmium sulphide, in a range of yellow shades, and cadmium reds which are co-precipitates of cadmium sulphide and cadmium selenides. Large quantities of iron-oxide pigments are now made synthetically in shades of yellow, red, brown, and black and, being manufactured products, are relatively pure and free from the silica and clay associated with the natural product.

As regards organic pigments, it will be obvious from the preceding sections that a very wide range of shade can be obtained by varying the chemical structure of the product in the same manner as with dyestuffs, the presence or absence of groups conferring the property of solubility having little effect on shade. Organic pigments fall into two groups. The first consist of pigments which are insoluble because they contain no chemical groupings which confer solubility on the molecule. The second group consists of pigments made from soluble dyestuffs by precipitation on a carrier which makes them insoluble. In the latter case a proportion of white base is an essential component and such pigments are commonly referred to as lakes. An example of this group is Madder Lake or Alizarine Lake, which is the insoluble pigment formed from the dyestuff alizarin. It is comparable with the colouring matter formed in the dyeing of textiles by the same dyestuffs. It will be readily understood that organic pigments, consisting as they do of compounds having different chemical structures, have a wide range of properties. Fastness to light, to acid and alkali, to heat, and to 'bleeding' in oils and solvents are properties which must be taken into account when choosing a pigment for a particular purpose. The pigments which have been developed commercially over the last twenty years are the results of a search for chemical combinations having better properties than those formerly available. One example which may be mentioned is Monastral Fast Blue BS, which is highly resistant to many chemical reagents and is in general characterized by great stability. Chemically it is akin to the natural pigment of the blood. Although the fastness properties already indicated depend on the chemical nature of the compound, pigments are in general more dependent for their satisfactory use on their physical condition and properties than are dyes. This is because whereas textile dyestuffs are dissolved at some stage in the process of dyeing, pigments are used as insoluble particles; the colour and many other properties depend on the size of the individual particles of pigment and the way in which they clump together.

Since pigments are insoluble and lack any dyeing power they can be used in two ways only. The first is by application to a surface which it is desired to colour; in this case the pigment particle must be held in position on the surface by some binder or adhesive, as in protective or decorative paints. Similar examples are the making of printing inks of various types, wallpaper printing, the coating of paper, and so on. In the second alternative the pigment may be mixed throughout the material to be coloured, which itself acts as the binder. For example, pigments are added during the compounding of the ingredients of rubber or of plastics, and when finally cured or moulded the finished article is coloured throughout its bulk. Other examples are the use of pigments in the manufacture of inlaid linoleum, and in the mass pigmentation of rayon. In the latter case a dispersion of a pigment in viscose is extruded in filament form, resulting in a viscose rayon which is coloured throughout each fibre at the time of its formation.

As already indicated, paint comprises a pigment, either white or coloured, dispersed in a binder. This binder may be a drying oil, such as linseed oil, or one of the newer synthetic resins. Such a mixture would be impossible to apply by any convenient means, but by adding a thinner or solvent the resulting compositions can be brushed or sprayed. Lacquers for motor cars are largely based on nitrocellulose as a binder. Many common household articles of metal are coloured with an enamel based on a synthetic resin which can be hardened by stoving. In all of these processes pigments are used for decorative effects.

Allied to paints are printing inks, of which there are several varieties, the choice depending on the kind of machines used for printing. For letterpress and lithographic printing the medium is usually linseed oil with a high proportion of pigment, the medium drying largely by oxidation by the air. In rotogravure inks, a solid resin is dissolved in a quick-drying solvent; these inks dry by evaporation. Newspaper inks, on the other hand, where cheapness is an important factor, generally consist of a mineral oil containing some resin in solution; these dry by absorption into the paper. It is possible to reproduce paintings and other illustrations in colour by three or four printings with inks with selected colours, whereby a complete range of tones can be reproduced. For this purpose the selection of pigments for the inks is highly critical and the manufacture of this type of ink is a very skilled business.

Certain metals, such as aluminium and copper, are used in the form of powders as pigments for paint and for other purposes. Paints containing very high proportions of zinc dust have recently been introduced to give anti-corrosive coatings similar to those produced by galvanizing. Finely divided carbon, such as Carbon Black, is a widely used black pigment, especially for motor-car tyres and printers' ink.

The use of pigments can also be seen in leathercloth, cosmetics, and concrete, in addition to those outlets already indicated. Naturally the properties required may differ considerably for these different uses but by their use it is possible to provide colour in many formerly drab aspects of everyday life.

CHAPTER FOURTEEN

Explosives

As was to be expected, the war of 1914–18 not merely enormously increased the demand for explosives of all kinds but effected certain changes and improvements. The manufacture of picric acid – a nitrated form of carbolic acid – was at first greatly increased but later fell as new explosives called amatols began to be introduced; these are mixtures of ammonium nitrate and trinitrotoluene (TNT). TNT is to-day the most widely used of all military explosives because it is cheaply and easily manufactured.

As interest in high explosives is primarily, but by no means exclusively, military it is perhaps appropriate to discuss their wartime uses before passing on to consider other applications. Nitro-glycerine, nitrocellulose, trinitrotoluene, and picric acid have already been discussed in considering the contribution made by the chemical industry in the war of 1914–18. Until recently glycerine was derived solely as a by-product of soap manufacture, but synthetic glycerine is now being manufactured on a large scale in the United States. Nitro-glycerine is a very powerful explosive but it is not entirely satisfactory in very cold climates; for this reason nitro derivatives of chemical relatives of glycerine, such as glycol, are being investigated. Various kinds of nitrocellulose are made by altering the extent of the treatment with nitric acid; the term gun-cotton is reserved for those containing more than 13 per cent of nitrogen. Nitrocellulose is, incidentally, used also for making certain types of lacquers and for the manufacture of substitutes for leather.

A number of other explosives based on the treatment of organic substances with nitric acid are in general military use; the development of these was, of course, greatly stimulated by the war of 1939–45. Pentaerythritol tetranitrate (PETN) is of interest in that it is wholly synthetic, Hexogen (Cyclonite, T4, RDX) also is synthetic, being made by the action of nitric acid on a substance, known as cyclotrimethylene trinitramine, which is obtained from ammonia and formaldehyde.

Mixtures of explosives of the nitro type are often used. Torpex, for example, is a mixture of RDX and TNT; it is used particularly for filling the war-heads of torpedoes. Pentolite is a mixture of PETN and TNT.

It will be noted that all high explosives are compounds of the element nitrogen. The ultimate source of this nitrogen is the air, but it is introduced in the form of nitric acid derived from synthetic ammonia. Nitric acid, whose importance in making dyes, drugs, and other organic chemicals has already been stressed, is thus also a vital material for the making of every type of modern explosive. Sulphuric acid, which is used for nitration in conjunction with nitric acid, is no less important.

For military purposes explosives have two main uses; one is to inflict damage on the enemy and his fortifications, the other is to deliver the destructive charges or bullets. For the latter purpose propellants are needed. From these less violent results are required; they must explode uniformly and comparatively slowly so that the projectile leaves the firearm smoothly and at a predetermined velocity. Propellants are often termed 'powders', but in fact they differ considerably in their properties. Cordite, for example, as its name implies, was originally made up in long cords. Ballistite is used as a propellant in the form of flakes. Propellants for rifle and shotgun cartridges are fine powders; they are based on nitrocellulose but many additions are made to give such qualities as low flash and smoke, and to eliminate flash-back.

A number of chemicals with explosive properties are used to initiate the detonation of high explosives; many of the latter are unaffected even by violent blows and will burn like a candle. Mercury fulminate, Nobel's original detonator, is still widely used. Certain other salts, such as lead azide and styphnate and silver azide, are also used. A range of complex nitrogenous compounds, such as tetrazene and trinitrotriazidobenzene, are also employed as detonators. They are made in comparatively small quantities since even one gram (one-thirtieth of an ounce) may be sufficient to detonate a very large charge of high explosive. The detonator itself may be set off mechanically, as by the firing-pin of a rifle, or electrically. Sometimes a secondary charge, known as a primer, is interposed between the detonator and the main charge; for this purpose tetryl is now commonly used.

In order to delay explosion, so as to protect the operator or for other reasons, fuses are often used. These consist of long tubes, generally of cotton made waterproof with rubber, filled with gunpowder. They are uniformly packed so that they will burn at a constant rate; a length can thus be cut appropriate to the delay required. What are called 'instantaneous' fuses are also made; these are packed with high explosive such as PETN. They burn at the rate of about four miles per second and are used if a number of charges have to be fired simultaneously, for example in demolition work.

Besides their military uses, high explosives find innumerable applications for peaceful purposes, since they provide a quick and easy means of moving and breaking up large masses of rock, soil, masonry, and so on. Many of the world's greatest feats of civil engineering – such as the Khyber railway and the Simplon tunnel – would have been quite impossible without the extensive use of blasting explosives. Thus in 1938 some 21,000 tons of explosives were used in Britain, a great part of it for blasting in coal mines; for this purpose alone 63 million shots were fired. Much is used in other mines and in quarries. It is interesting to reflect that the explosives produced by the chemical industry are valuable – indeed indispensable – in providing many of the raw materials, such as coal, limestone, and anhydrite, which it requires in great quantities.

Blasting – except in coal mines where there is a fire hazard – is done mostly with blackpowder, which differs very little from the original gunpowder of the thirteenth century. It consist of a mixture of potassium nitrate, sulphur, and charcoal. Its use for blasting purposes dates from the seventeenth century. It owes its explosive properties to its power of rapid combustion, by which a large volume of gas is generated in a very short space of time, but for this reason cannot be used in mines where fire-damp may be encountered; in Britain special explosives have been developed for use in 'fiery' mines of this kind and their use is now obligatory. Despite this limitation about 3,500 tons of blackpowder are made annually in Britain, mostly for blasting purposes.

Another important peacetime use of explosives is in pyrotechnics. Apart from the making of many kinds of fireworks for display purposes, small quantities of explosives go into life-saving rockets, distress flares, Very lights, self-igniting floats, and similar appliances.

CHAPTER FIFTEEN

Plastics

ALTHOUGH this branch of the chemical industry is the one with which the general public most frequently comes into direct contact it is nevertheless one about which many misconceptions exist. Plastics are often spoken of as though there was little difference between the various kinds; in fact they differ enormously in their properties. Plastics are often thought of as new substances; in fact they have been in use for a century. Plastics are often regarded as cheap substitutes for other and better constructional materials such as wood, metal, and natural textiles; in fact many have found favour on their own merits and often are far from cheap.

Plastics may be defined as substances which, at some stage or another, can be made to flow and which can be fixed permanently in any shape they have been made to assume. Sometimes they are liquids and flow easily; sometimes they form stiff pastes and can only be shaped by the application of pressure. Their many practical applications arise largely from the fact that they can be shaped in moulds and thus lend themselves to the comparatively cheap mass-production of objects of intricate shape. Certain of them may be formed into fibres which, woven into cloth, form valuable supplements to natural textiles such as cotton, wool, and silk.

Chemically, plastics differ enormously in their constitution and method of manufacture. Nevertheless they have one common factor; all consist of giant molecules. Most of the substances which we have so far considered have molecules – by which are meant the smallest units of the substance which can exist and display all its characteristic properties – made up of, at the most, a few score atoms. The architecture of plastics is quite different; their molecules contain tens or hundreds of thousands of atoms. These giant molecules possess, however, a regular structure, for they consist of the almost endless repetition of comparatively simple atomic patterns. Because of this repetitive structure, which has been clearly revealed by X-ray analysis and in other ways, substances consisting of these giant molecules are known as polymers, a word of Greek derivation, meaning literally 'many parts'. The molecules of polymers may be likened to chains each consisting of many identical links.

Although many modern plastics are of purely synthetic origin many natural polymers exist and it is not unnatural that it was to these that the chemist first directed his attention. Cellulose is an important example of a natural polymer; its importance in the explosives industry was referred to in the last chapter. When suitably prepared, nitrocellulose is a powerful explosive; rather surprisingly it is also an ingredient of one of the oldest – but still widely used – plastics. At about the middle of the last century Alexander Parkes, a Birmingham man, obtained a solid product, which he named Xylonite, by mixing nitrocellulose, camphor, and castor-oil. His early products were not entirely satisfactory and it fell to an American, Hyatt, to make the first commercially useful product, which was named celluloid; this was a mixture of nitrocellulose and camphor, and it was first manufactured as long ago as 1872.

Other chemists investigated cellulose in a different way. John Mercer – who discovered that cotton could be made lustrous by treatment with caustic soda – found that cellulose could be dissolved in a solution of copper oxide in ammonia. A French chemist, Schützenberger, discovered another solvent for cellulose, acetic acid. Cheap solvents were also discovered for nitrocellulose. The first practical result of these discoveries was the development of lacquers such as those now so widely used in the motor industry. When the solvent evaporates from these solutions the cellulose or nitrocellulose is left as a hard glossy film. If deposited on fabric, artificial leather – or leathercloth – can be made. Thin films of cellulose acetate are transparent and flexible and are very suitable for wrapping food, cigarettes, and fancy goods.

Cellulose is, however, put to an even more important use by the modern chemical industry – fulfilling a prophecy made as long ago as 1664 by Hooke:

I have often thought that probably there might be a way found out, to make an artificial glutinous composition much resembling . . . that Excrement . . . out of which the Silkworm wire-draws his clew. If such a composition were found, it were certainly an easy matter to find very quick ways of drawing it out into small wires for use. I need not mention the use of such an Invention, nor the benefit that is likely to accrue to the finder.

In 1855 a crude artificial silk was made in France by treating mulberry leaves with nitric acid and making the resulting crude nitrocellulose sticky by addition of rubber. Soon afterwards, however, a new incentive to experiment arose, for the immense expansion in newspaper production made wood-pulp – a comparatively pure form of cellulose – cheaply and abundantly available. In 1883 Sir Joseph Swan made the first practical artificial silk by squirting a solution of nitrocellulose in acetic acid into a bath of alcohol, which coagulated the jets to threads which he originally used, after ignition, as filaments for the newly discovered electric lamps. Later, as a result of collaboration with L. S. Powell, who had suggested the use of zinc chloride for dissolving cellulose, Swan produced much finer threads and some of these were crocheted by his wife into articles which were shown at the Inventions Exhibition of 1885 under the name 'artificial silk'. Swan did not pursue his discovery commercially, and the first to manufacture artificial silk was the Count de Chardonnet whose factory at Besançon was in operation by 1901; by 1907 he was making a ton a day. In Britain, meanwhile, interest was still alive and in 1892 Cross, Bevan, and Beadle developed the very important viscose process in which alkali-treated cellulose is dissolved in carbon disulphide. This process was economically very important, as it required only the cheap cellulose of wood-pulp and not expensive pure cellulose such as cotton. Courtaulds began the industrial operation of the viscose process in 1906 and to-day, in factories in many different countries, it provides three-quarters of the world's artificial silk. Much of the remainder is what is called acetate silk – or Celanese – made by a process perfected by Dreyfus during the First World War. The immense industrial importance of rayon to-day may be measured by the fact that world production is about a million tons a year. About a half of this is made in the United States; the next largest producer is Great Britain.

Other artificial fibres are of great importance in the textile industry. The best known of these is nylon, an

American discovery used primarily for the hosiery industry but finding many other uses as well, for example in making underwear and other clothing, ropes, nets, fishing lines, bristles, and solid mouldings. Nylon filament has a silky appearance, is resistant to dampness and mildew, and is more elastic and stronger than silk. The provision of the new materials necessary for making nylon on a large scale greatly taxed the ingenuity of the chemical industry.

Recently nylon has been joined by a very important British fibre – of a chemically different nature – called Terylene; this is being manufactured also in the United States under the name Dacron. Like nylon, Terylene can be made either as a continuous fibre or as staple fibre which can be spun. Its full field of application has yet to be realized, but will include the making of both light and heavy fabrics, trawl-nets, fishing lines, ropes, and hosiery. It is one of the strongest fibres known and is resistant to light, mildew, and many chemicals. The principal raw materials for the manufacture of Terylene are terephthalic acid – derived from petroleum – and ethylene glycol, a syrupy liquid familiar to all motorists as anti-freeze for radiators. The Terylene project represents an expenditure on research and development of some £(m)15.

Other important fibres provided by chemists for the textile industry are derived from natural proteins. Silk itself is a protein and its fibrous form is a reflection of its molecular architecture, for it consists of long, narrow molecules arranged parallel to each other. Many other proteins, such as the casein of milk or the protein of ground-nuts, have similar molecules with the important exception that instead of being extended and aligned they are twisted up into balls. By chemical treatment, however, it is possible to straighten out these molecules and thus obtain fibrous proteins. Outstanding among these products is the British fibre Ardil, derived from ground-nut protein. It has some resemblance to wool, with which it may be blended, and can be mixed with rayon or other textiles to give a new range of fabrics.

A natural plastic of the utmost importance is rubber, the importance of which in the modern world needs no emphasis. Production of natural rubber rose, under the urgent needs of war, to no less than $1\frac{1}{2}$ million tons in 1941. Then, as a result of the Japanese occupation of the principal producers, the United Nations were able to obtain no more than one-sixth of this quantity. But for the truly remarkable effort made by the American chemical industry in providing, in an almost unbelievably short space of time, huge quantities of synthetic rubber of excellent quality the results of this loss of natural rubber would have been catastrophic. A variety of synthetic rubbers were made, each resembling natural rubber in its general chemical architecture. The most important type was GR-S used particularly for tyres; in 1945 production of this reached a peak of three-quarters of a million tons. Butyl rubber was made for inner tubes. Neoprene was made for special purposes in which high resistance to oil and solvents was important. Germany had, of course, well-prepared plans for rubber synthesis and also made large quantities of synthetic material, sufficient to meet her essential needs. Russia, too, developed a limited synthetic rubber industry.

Plastics may for convenience be divided into two main classes – thermo-setting and thermo-softening. Thermo-setting plastics are those in which the final form is made permanent by the application of heat; this brings about irreversible chemical changes and no further change in shape is possible except by such mechanical means as grinding, chipping, sawing, and so on. Thermo-softening plastics are those which, like sealing-wax, become soft and malleable on heating but become hard again on cooling, the process being capable of repetition as often as desired. Akin to the thermo-setting plastics are those in which hardening is permanently effected by the addition of some substance which brings about a chemical change; sometimes this can be made to occur at ordinary temperatures.

Among the most important of the thermo-setting plastics are those made from phenol – much of which is now made synthetically from benzene – and formaldehyde. Frequently cresol, a cheaper relative of phenol, is used; this gives a darker product but for many purposes this is not a serious disadvantage. Plastics of this type were pioneered by Baekeland and are often known collectively as Bakelites. They are supplied by the chemical industry in the form of powders, mixed with suitable fillers, such as wood-flour, and used for making a very wide range of mouldings. They are also used as decorative finishes, for bonding laminated materials such as plywood, and for impregnating paper and fabrics. Enormous quantities are made. In January 1948, for example, the United States alone manufactured almost 30 million pounds of Bakelite plastics.

Because of their dark colour Bakelite plastics do not lend themselves to coloration with pigments. This objection does not apply, however, to what are called Beetle plastics, a type made by chemical reaction between synthetic urea – which in nature occurs as a principal constituent of urine – and formaldehyde. Various modifications of this fundamental chemical reaction are employed. In place of urea may be used related substances such as melamine or thiourea; formaldehyde may be replaced by other aldehydes. By introducing suitable pigments into the resulting moulding powders very brightly coloured products are obtainable.

The use of casein and other natural products for the making of artificial textiles has been referred to; they may also be used for making plastics for moulding processes. Proteins from milk, soya beans, and coffee beans have been used for this purpose. The process is usually a complicated one, involving prolonged treatment of the protein with formaldehyde and much mechanical working. The products may be brightly coloured with pigments.

The alkyd plastics are made by chemical reaction between substances such as glycerine or ethylene glycol and citric or phthalic acids. They are very extensively used as synthetic finishes and for bonding other materials, particularly in the United States.

Comparatively new arrivals in the thermo-setting class of plastics are the allyl resins. These are derived from a substance known as allyl alcohol, made from petroleum. They are supplied as yellowish treacly liquids which set to an amber-coloured glass when a small quantity of an accelerating substance is added. A notable feature of this solidification is that it occurs without change of volume, without need to apply pressure, and without release of any by-product such as water or ammonia. Their manipulation is therefore possible with extremely simple equipment.

In recent years a wide range of thermo-setting plastics have been used to replace glues of traditional types, especially in preparing plywood and other laminated materials. Some of them are stronger and more durable than wood and they have the advantage of resisting water, dry heat, and the attacks of fungi.

Outstanding among the thermo-softening plastics – materials which, if carefully handled, can be melted and resolidified as often as desired – is polythene. This is a tough, waxy-looking substance with remarkable powers of electrical insulation. A wholly British discovery, first produced industrially on the very day on which war broke out in 1939, it played a vital role in the development of radar. It is a water-resistant plastic and in the form of sheet it is used for such diverse purposes as wrapping drugs, making waterproof covers for motor cars during shipment overseas, and lining industrial containers. Polythene tubing is widely used for carrying water on farms, and is increasingly replacing lead, at present exceedingly expensive, for domestic and industrial plumbing. A chemical relative – Fluon – in which the hydrogen of polythene is replaced by fluorine has even more remarkable properties; it will withstand great heat and the action of almost all known chemicals, including such powerfully corrosive ones as molten caustic soda. Fluon is, however, at present a costly material and its uses will therefore be limited for the time being to applications for which normal materials are unsuitable; it is, for example, already finding use in making bases for radio valves and for making heat-resisting gaskets.

Akin to polythene is polyvinyl chloride (PVC); this is a polymerized form of ethylene in which part of the hydrogen has been replaced by chlorine. The product is a tough leathery material much used in the form of sheeting, for example for making handbags and other articles normally made from leather, tobacco-pouches, tablecloths, and raincoats. It is also much used for making insulating covers for electric cables and similar purposes. It is used in surgery for making artificial noses, ears, and so on. PVC is also made in fibrous form and worked up into heavy fabric for making protective clothing, filter cloth, and other industrial materials. The starting materials for its manufacture are acetylene and hydrochloric acid.

If ethylene and benzene are combined into one molecule and this is then polymerized the product is the transparent plastic polystyrene; this is much used as an unbreakable substitute for glass, for example in making tumblers, spoons, meat-covers, wall-tiles, and other domestic articles. The starting material – styrene – is a product of the petroleum industry.

Another extremely important transparent plastic is that known as Perspex and by other trade names. The principal raw material is acetone; chemically it is a polymerized form of a liquid known as methyl methacrylate. By somewhat varying the chemical architecture the properties of the plastic, for example its power of transmitting light and temperature of softening, can be varied. Perspex finds very many uses. After softening, sheets of Perspex can, for example, be pressed into large transparent domes for aircraft. Lenses of good optical quality can be cast in Perspex; this development shows some promise for making large compound lenses, which are very costly to make from glass. Most artificial dentures are to-day made from Perspex. The medical profession uses it for making splints because, as it is so easily shaped after softening by heat, it can be made to fit according to the individual requirements of the patient.

Finally, particular mention must be made of a new class of plastics – the silicones – which are of great technical importance. They differ fundamentally from all the plastics previously discussed in that the principal units of the polymer molecules are not carbon atoms but atoms of silicon. The latter is very well known in the form of its oxide, which is the main ingredient of sand.

The silicones assume many forms; some are thin liquids, some treacly, some putty-like, some rubbery, some solid. They have the very valuable common property of great resistance to extremes of temperature and a capacity to repel water. They are particularly useful, in view of their water-repelling power, as electrical insulators in exposed positions. Greases based on silicones neither stiffen at low temperatures nor char on moderate heating. Liquid silicones do not become viscous on cooling and have been used in the hydraulic systems of aircraft.

At the present time, however, the full value of silicones has yet to be discovered as they are still expensive and accordingly restricted to applications in which their unique properties are of exceptional value.

The present range of application of plastics is immense. In the electrical industry they are paramount, for they alone provide the means of mass-producing the intricately shaped insulating parts essential for all modern electrical appliances. Most motor cars contain dozens of plastic components. Every home contains numerous plastic articles. Their present significance may be judged by the following table which summarizes the production of plastics in Britain in recent years.

Type of Material	Production 1941 (tons)	Production 1946 (tons)
Cellulose acetate	3,452	5,737
Celluloid	1,719	2,100
Perspex	1,463	3,949
PVC	—	5,471
Phenol-formaldehyde	13,521	25,221
Urea-formaldehyde	4,136	6,471
Total:	24,291	48,949

This considerable total is, however, far outstripped by that for the United States, which is now the world's biggest producer and user of plastics and in 1947 manufactured a total of 370,000 tons.

Many plastics are comparatively costly; this is often obscured by the fact that plastic articles are often surprisingly cheap. Several factors contribute to this apparent paradox. Plastics have a low specific gravity, and thus a smaller weight of material is needed to make an object of given size than if, for example, a light metal alloy were used. Again, much of the cost of an article made in say wood or metal represents labour; plastics, because they lend themselves so admirably to mass production, require comparatively little labour per article.

The working of plastics involves much skill and the use of very complicated mechanical equipment, but it is the chemical industry alone which provides the raw materials in use to-day and from whose researches spring the steady flow of new plastics and ancillary chemicals. In this study of the chemical industry plastics make a special claim on our attention not because

they represent an exceptionally large part of the industry – in either quantity or value – but because we
encounter them constantly in all their different forms in
daily life; in a sense they epitomize the diversity of the
products of the industry as a whole.

PART III: FUTURE

CHAPTER SIXTEEN

The Future of the Chemical Industry

HAVING surveyed the history of the chemical industry and discussed its present structure, it is tempting to go on to speculate about its future. For so complex an organization, closely linked to almost all other branches of industry and thus sensitive to changes far outside its own orbit, prophecy is an even more dangerous pastime than usual. Industry as a whole is affected by many factors – such as the threat of war, the promise of lasting peace, and political activity in distant countries – and the chemical industry must constantly adapt itself to all these fluctuations. At the best, therefore, we can discuss its future only in very general terms.

One prediction can safely be made; it is that the chemical industry will expand to keep pace with the expansion of industry generally in Europe and the American continent. In the many backward countries which are striving to obtain a higher standard of living by the well-tried road of industrial development, chemical industries must be created. As far ahead as one can foresee there will, therefore, be a steadily increasing demand for all the basic chemicals such as sulphuric and nitric acids, alkali, and ammonia. In addition one can reasonably foresee a great expansion in certain other existing branches. Although plastics, for example, are already widely applied there seems no doubt that they are very far from being fully exploited. Existing plastics will find new uses and entirely new ones will be developed meeting needs at present unfulfilled. The production of synthetic textiles, already very great, offers exceptionally great opportunities to the chemist. To the artificial fibres based on cellulose, and to nylon and the newcomer Terylene, will certainly be added others with no less remarkable properties.

Another general prediction which can safely be made is that the chemical industry will have to adjust itself to utilizing extensively a number of raw materials different from those to which it is now accustomed. The recent critical position with regard to sulphur has been referred to in an earlier chapter. Pure sulphur, mined in immense quantities in Texas and on a very much smaller scale in Sicily, is, as we have seen, the simplest raw material for the manufacture of sulphuric acid; unfortunately, however, the American deposits must inevitably be exhausted by the huge inroads now being made into them and cannot at the present time easily be exploited as rapidly as the increasing demands of industry require. In the United States itself, and to a far greater extent in the many countries which have become accustomed to importing American sulphur, increasing use must be made of various forms of combined sulphur such as anhydrite and pyrites. The rapid and accelerating growth of the organic chemical industry, too, poses many problems in regard to raw materials. Coal, originally almost the sole source of carbon for industrial organic chemicals, has been joined by petroleum; it is quite certain that the next few years will see a tremendous expansion in the use of petroleum by the chemical industry.

New manufacturing techniques will certainly make their appearance. In particular, increasing use is likely to be made of high pressures, by which effects can be produced not obtainable in other ways. In this connexion it is interesting to reflect that even at the beginning of this century the idea of manufacturing tens of thousands of tons of chemicals by the use of pressures hundreds of times greater than those of the atmosphere, and at high temperatures, would have seemed fantastic. Yet to-day manufacturing under these conditions of pressure and temperature is commonplace all over the world; in the future much higher pressures may be used. At ten thousand atmospheres a million cubic feet of air, weighing about thirty-five tons, could be compressed into less than a five foot cube. Laboratory experiments, notably those of Bridgman in the United States, have shown that at pressures of this order many common substances undergo remarkable changes and industrial exploitation of these effects may have most interesting and useful results, leading to entirely new products.

Of immediate international interest is the remarkable resurgence of the German chemical industry, which in 1946 was virtually non-existent. To-day, however, West Germany is not merely meeting her needs but is once again a formidable competitor for exports. In 1938 German chemical exports represented $(m)300; in 1951 they were $(m)503, or 12 per cent of the world's export chemical sales – very nearly as great as Britain's share. In 1949–50 Germany had to import 40,000 tons of nitrogenous fertilizers; in 1951 she had an export surplus of 150,000 tons. More remarkable still, Germany is again the greatest dye exporter, having one-third of the world's export trade. In drug export, too, Germany has recovered, now having 8 per cent of the world's trade, compared with 40 per cent before World War II.

In an earlier chapter it was shown that the alchemists' dream of transmuting base metals into gold was – although an idle fancy – responsible for the discovery of much new chemical knowledge. It is interesting to reflect that, by methods of which the alchemists had no knowledge, the transmutation of metals has not merely been achieved in the laboratory but is to-day carried out as an essentially industrial operation. The quantities produced are admittedly extremely small, but their value for research is immense, and has enabled the frontiers of science to be moved forward in many different directions. From Britain's Atomic Energy Research Establishment at Harwell, for example, these artificial elements are sent to laboratories throughout Europe, in South Africa, and even farther afield. Other large production centres exist in the United States, France, and elsewhere. The great importance of these radio-elements for research and for medical diagnosis and treatment will undoubtedly lead to their production on a much greater scale than at present, although the actual weights produced will be very small, both for technical reasons and because their high radioactivity enables important results to be obtained with no more than traces.

With this brief account of the realization of the alchemists' ardent but unfulfilled dreams we may appropriately end the story of the chemical industry. Undoubtedly in the future it will achieve results as remarkable, but to-day as apparently unattainable, as transmutation was to the alchemists. Whatever these new achievements may be we can be certain that they will ultimately be reflected, directly or indirectly, in daily life.

Reading List

Some Founders of the Chemical Industry, by J. Fenwick Allen. Sherratt and Hughes, 1906.

The First Fifty Years of Brunner, Mond and Company. Brunner, Mond and Co, 1923.

History of the British Chemical Industry, by Stephen Miall. Ernest Benn Ltd, 1931.

A Hundred Years of Chemistry, by Alexander Findlay. Duckworth, 1937.

Thorpe's Dictionary of Applied Chemistry. Longmans, Green and Co, Ltd, 4th edition, 1937.

British Chemical Industry; its Rise and Development, by G. T. Morgan and D. D. Pratt. Edward Arnold, 1938.

Chemistry in the Service of Man, by Alexander Findlay. Longmans, Green and Co, Ltd, 7th ed., 1947.

Fifty Years of Progress; the Story of the Castner-Kellner Alkali Company. Imperial Chemical Industries Ltd, 1947.

A History of Chemistry in Canada, compiled by C. J. S. Warrington and R. V. V. Nicholls. Sir Isaac Pitman and Sons (Canada) Ltd, 1949.

Report on the Chemical Industry (1949). Association of British Chemical Manufacturers, 1949.

A History of the Chemical Industry in Widnes, by D. W. F. Hardie. Imperial Chemical Industries Ltd, 1950.

Minerals in Industry, by W. R. Jones. Penguin Books. 2nd ed., 1950.

A Century of Science, edited by R. Partington. Hutchinsons Scientific and Technical Publications, 1951.

A Century of Technology, edited by Percy Dunsheath. Hutchinsons Scientific and Technical Publications, 1951.

Traité de Chimie industrielle, by Paul Baud. Masson et Cie, 4th ed., 1951.

Index